Artificial Intelligence in the Production of Biotherapeutics

The transformative role of artificial intelligence (AI) in modern biomanufacturing focuses on key areas such as process analytical technology (PAT), Good Manufacturing Practice (GMP) compliance, predictive analytics, and AI-driven quality systems. It bridges cutting-edge AI applications with the complexities of biotherapeutic production, offering insights into automation, real-time monitoring, and process optimization.

Delving into the core of biomanufacturing, the book provides a structured journey through its critical phases. It begins with an introduction to modern biomanufacturing principles, quality by design approaches, and the integration of AI. Subsequent chapters examine raw material management, lean manufacturing practices, and the application of predictive analytics to optimize supply chains. Readers will explore advanced tools such as AI-enhanced data acquisition in PAT, automated standard operating procedures (SOPs), and AI-driven process controls for fermenters and chromatography systems. The text also addresses GMP essentials, including personnel management, hygienic facility design, and pharmaceutical water systems. Key chapters highlight AI's role in validation processes, sterile packaging, and regulatory compliance, referencing global guidelines from organizations such as the WHO, FDA, and ICH. Real-world case studies featuring therapeutic proteins, monoclonal antibodies, and vaccines underscore the practical applications of AI in scaling and maintaining biotherapeutic production.

This book equips readers with a comprehensive understanding of AI's potential to enhance efficiency, accuracy, and compliance in biomanufacturing. Whether you are a professional, researcher, or student, this guide offers actionable insights into leveraging AI to revolutionize biotherapeutic production while adhering to the highest industry standards.

What You Will Learn:

- Understand how AI enhances every phase of biotherapeutic production, from raw material management to regulatory compliance, optimizing efficiency, accuracy, and quality.
- Explore the role of AI in advanced data acquisition, process control, and continuous improvement, including applications in fermenters, flow filtration, and chromatography systems.
- Gain insights into leveraging AI for automating standard operating procedures (SOPs), predictive maintenance, quality assurance, and adhering to global GMP standards like WHO and FDA guidelines.
- Learn how AI transforms upstream and downstream processes, ensures sterility in packaging, and supports case studies on therapeutic proteins, monoclonal antibodies, and human vaccines.
- Discover the potential of AI in shaping the future of biomanufacturing, including challenges, data security, and the ethical implications of AI-driven automation.

Artificial Intelligence in the Production of Biotherapeutics

Principles, Practices, and Standards

Alok Kumar Srivastav
Priyanka Das

CRC Press
Taylor & Francis Group
Boca Raton London New York

CRC Press is an imprint of the
Taylor & Francis Group, an **informa** business

First edition published 2026
by CRC Press
2385 NW Executive Center Drive, Suite 320, Boca Raton FL 33431

and by CRC Press
4 Park Square, Milton Park, Abingdon, Oxon, OX14 4RN

CRC Press is an imprint of Taylor & Francis Group, LLC

ISBN: 978-1-041-03635-7 (hbk)
ISBN: 978-1-041-03637-1 (pbk)
ISBN: 978-1-003-62459-2 (ebk)

DOI 10.1201/ 9781003624592

Typeset in Times
By Deanta Global Publishing Services, Chennai, India

To my family and friends, whose unwavering support and encouragement have been my guiding light throughout this journey. Your belief in me has been the foundation upon which this work stands.

To my students, whose curiosity and enthusiasm inspire every page I write.

And to those who seek to understand the world through both science and faith, may this book contribute to your journey of discovery and wonder.

–Dr. Alok Kumar Srivastav

Contents

Preface

The biomanufacturing sector has witnessed tremendous growth over the past few decades, driven by advancements in biotechnology, medicine, and technology. As the production of biotherapeutics – ranging from therapeutic proteins and monoclonal antibodies to vaccines and gene therapies – becomes more complex, there is a growing need for innovative solutions to streamline processes, improve efficiency, and maintain the highest standards of quality. One such innovation is the integration of artificial intelligence (AI) into biomanufacturing practices, which has the potential to revolutionize the industry.

This book, *Artificial Intelligence in the Production of Biotherapeutics: Principles, Practices, and Standards*, explores the transformative role that AI plays in modern biomanufacturing. From raw material management and process optimization to facility maintenance and regulatory compliance, AI is enhancing the precision, speed, and cost-effectiveness of biomanufacturing processes. By automating routine tasks, optimizing supply chains, and offering real-time decision-making support, AI enables manufacturers to meet the ever-growing demand for biopharmaceuticals while adhering to stringent quality standards and regulatory guidelines.

In the chapters that follow, we delve into the core principles of biomanufacturing, including the quality by design (QbD) approach, the importance of standard operating procedures (SOPs), and the implementation of Good Manufacturing Practices (GMPs). We also address the specific technical challenges encountered in upstream and downstream processing, the qualification of facilities, and the production of sterile packaging. Central to these discussions is how AI can enhance every aspect of the biomanufacturing life cycle, from improving raw material supply chains to automating complex validation processes.

In particular, we explore how AI's application within areas such as process analytical technology (PAT), predictive analytics, and quality systems has reshaped the landscape of biomanufacturing. The integration of machine learning algorithms and AI tools is enabling manufacturers to monitor and control processes with unprecedented precision, ensuring the consistency, safety, and efficacy of biotherapeutic products.

As biomanufacturing continues to evolve, this book serves as both a comprehensive guide and a forward-looking resource for professionals, researchers, and stakeholders within the industry. Whether you are a biomanufacturing expert or an AI enthusiast seeking to understand the synergies between these fields, this book provides invaluable insights into how AI is shaping the future of biotherapeutic production.

We hope that by the end of this journey, you will not only gain a deep understanding of biomanufacturing and its intersection with AI but also be inspired to explore new possibilities in this rapidly evolving field.

Acknowledgment

Writing a book is a journey that often involves the support, encouragement, and contributions of many individuals and organizations. As we present this work, *Artificial Intelligence in the Production of Biotherapeutics: Principles, Practices, and Standards*, we would like to express our heartfelt gratitude to those who have made this endeavor possible.

First and foremost, we extend our sincere appreciation to our families for their unwavering support, patience, and understanding throughout the writing process. Your encouragement has been a constant source of motivation. We are deeply thankful to our colleagues and mentors whose guidance and expertise have enriched this book. Your insights have shaped our understanding of the *Production of Biotherapeutics,* particularly AI. We would like to acknowledge the contributions of the research institutions and libraries that provided access to valuable resources, making our research more comprehensive and thorough. Our gratitude goes to the reviewers and experts in the field who provided valuable feedback and constructive criticism, helping us refine the content and ensure its accuracy.

We extend our thanks to the publishing team, editors, and designers who have worked diligently to transform our manuscript into a published book. Last but not least, we are grateful to our readers, students, and fellow researchers who find value in this book. Your interest and engagement in the subject of *Artificial Intelligence in the Production of Biotherapeutics: Principles, Practices, and Standards* drive our commitment to promoting sustainable practices and environmental stewardship.

This book would not have been possible without the collective effort and support of these individuals and institutions. We humbly acknowledge your contributions and express our deepest appreciation.

About the Authors

Alok Kumar Srivastav is an accomplished Assistant Professor in the Department of Health Science at the University of the People, Pasadena, CA. His academic background includes a PhD, MTech, and MSc in Biotechnology, a Postdoctoral Fellowship (Research) in Biotechnology from Lincoln University College, Malaysia, and an MBA in Human Resource Management. Dr Srivastav is a distinguished figure in Academia and Research, honored with the "International Pride of Educationist Award" at AIT, Thailand, in 2022 for pioneering contributions to advancing education in the digital era, and recipient of a prestigious "Innovative Academic Researcher Award" at HULT, France, UK, in 2024 for his exceptional creativity, innovation, and impact in academic research.

Priyanka Das serves as Assistant Professor in the Department of Health Sciences at the University of the People in Pasadena, CA. She holds a PhD, MTech, and MSc in Biotechnology along with an MBA in Human Resource Management. Prior to her current position, she was a Postdoctoral Fellow (Research) in Biotechnology at Lincoln University College, Malaysia. Dr Priyanka Das is a dedicated scholar, contributing significantly to the field of biotechnology.

1 Introduction to Biomanufacturing

1.1 INTRODUCTION

The landscape of biomanufacturing has undergone a remarkable transformation over the past several decades, evolving from simple fermentation processes into sophisticated, highly controlled manufacturing systems that integrate advanced technologies with biological principles. This evolution represents not merely a technological advancement but also a fundamental shift in how we approach the production of biological products, from therapeutic proteins and vaccines to advanced cell therapies and tissue-engineered constructs. The contemporary biomanufacturing industry stands at the intersection of multiple disciplines, including molecular biology, process engineering, automation technology, and quality systems, creating a complex yet highly efficient framework for the production of biological products at a commercial scale.

The historical progression of biomanufacturing reveals a fascinating journey from traditional brewing and fermentation processes to today's state-of-the-art facilities equipped with advanced monitoring systems and automated control mechanisms. This transformation has been driven by several factors, including advances in our understanding of cellular and molecular biology, improvements in analytical technologies, and the development of more sophisticated process control systems. The early days of biomanufacturing, characterized by simple batch processes and limited process understanding, have given way to highly optimized continuous processing systems that leverage real-time monitoring and advanced control strategies to ensure consistent product quality. This evolution has been particularly significant in the biopharmaceutical sector, where the demands for product quality and regulatory compliance have driven continuous innovation in manufacturing technologies and processes.

The foundation of modern biomanufacturing rests upon a sophisticated understanding of biological systems and their behavior in industrial-scale processes. This understanding encompasses not only the fundamental aspects of cell growth and protein expression but also the complex interactions between various process parameters and their impact on product quality. The development of robust manufacturing processes requires careful consideration of numerous factors, including cell line stability, media composition, process parameters, equipment design, and facility requirements. These considerations become particularly critical when scaling up processes from laboratory scale to commercial production, where maintaining consistent product quality across different scales presents significant challenges (Figure 1.1). The integration of quality by design principles has provided a systematic

DOI: 10.1201/9781003624592-1

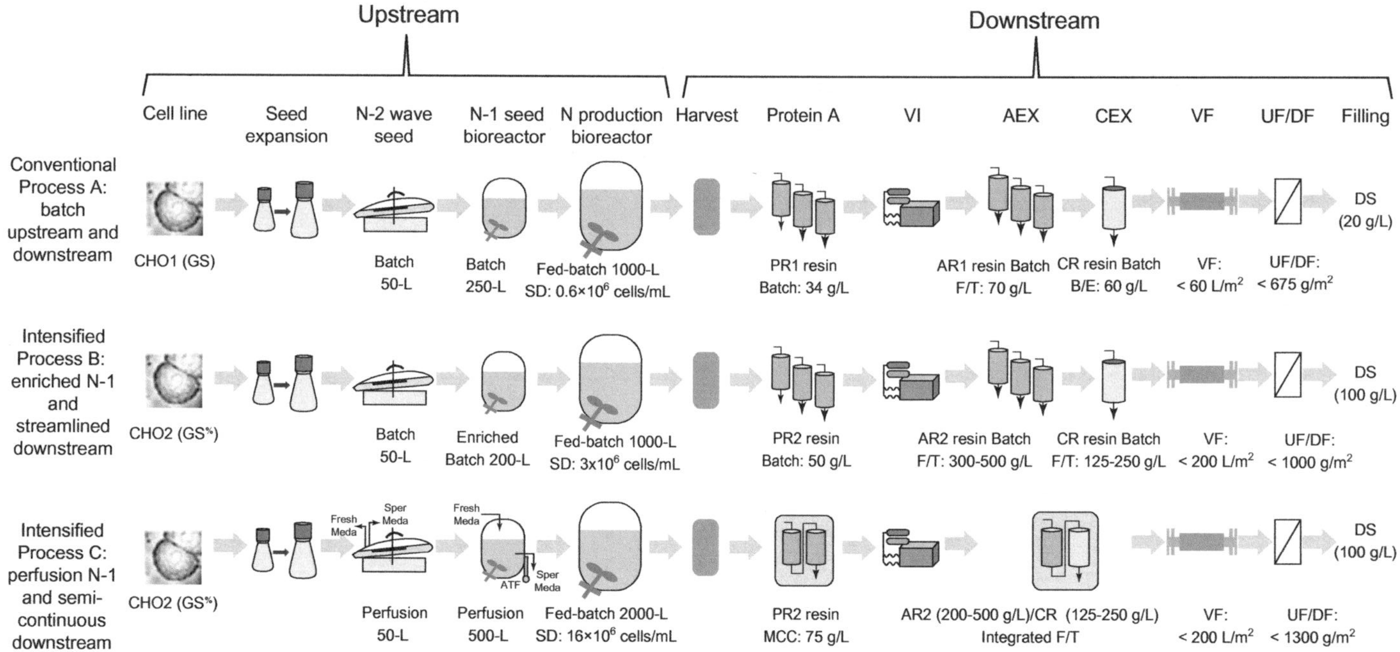

FIGURE 1.1 Evolution of biomanufacturing: From traditional processes to modern integrated systems.

framework for addressing these challenges, emphasizing the importance of understanding critical process parameters and their relationships to product quality attributes from the earliest stages of development.

The implementation of advanced process analytical technologies (PATs) has revolutionized how biomanufacturing processes are monitored and controlled. These technologies enable real-time measurement of critical process parameters and quality attributes, providing manufacturers with unprecedented insight into their processes. The ability to monitor and control processes in real time has led to significant improvements in process consistency and product quality while also enabling the development of more sophisticated control strategies. The integration of PATs with modern control systems has created a framework for adaptive process control, where manufacturing parameters can be adjusted in response to real-time measurements to maintain optimal process conditions and ensure consistent product quality.

The emergence of artificial intelligence (AI) and machine learning technologies represents the latest frontier in biomanufacturing innovation. These advanced analytical tools are being applied across the entire manufacturing life cycle, from process development and optimization to real-time process control and predictive maintenance. The ability to analyze large, complex datasets generated during manufacturing operations has enabled the identification of subtle patterns and relationships that might not be apparent through traditional analysis methods. This enhanced understanding has led to improvements in process efficiency and product quality while also providing new tools for process optimization and control. The integration of AI technologies with traditional manufacturing systems has created new opportunities for process improvement and optimization, pointing toward a future of increasingly sophisticated and efficient production systems.

The regulatory landscape surrounding biomanufacturing has also evolved significantly, with regulatory agencies adopting more flexible, risk-based approaches to manufacturing oversight. This evolution has been driven by the recognition that traditional regulatory frameworks, developed for small-molecule pharmaceutical manufacturing, may not be optimally suited to the unique challenges of biological product manufacturing. The implementation of quality by design principles and the adoption of advanced process control strategies have facilitated this evolution, providing regulators with greater confidence in manufacturers' ability to consistently produce high-quality products. This regulatory evolution has enabled the adoption of more innovative manufacturing approaches, including continuous processing and real-time release testing, while maintaining the high standards of product quality and safety required for biological products.

The future of biomanufacturing continues to evolve, driven by advances in technology and our growing understanding of biological systems. The integration of advanced analytics, artificial intelligence, and automated control systems is creating new opportunities for process optimization and improvement. These developments are particularly important as the industry faces new challenges, including the production of increasingly complex biological products and the need for more flexible, adaptable manufacturing systems. The continued evolution of biomanufacturing technologies and approaches will be critical in meeting these challenges while maintaining the high standards of product quality and safety required for biological products.

1.2 OVERVIEW OF MODERN BIOMANUFACTURING PRINCIPLES

The foundational principles of biomanufacturing encompass a complex interplay of biological processes, engineering controls, and regulatory requirements that collectively define the modern approach to biological product manufacturing. These principles have evolved significantly over the past several decades, driven by advances in our understanding of cellular biology, improvements in process control technologies, and the increasing demands for product quality and consistency. The integration of these various elements has created a sophisticated framework for biological product manufacturing that continues to evolve as new technologies and approaches emerge (Figure 1.2). Within this framework, manufacturers must carefully balance

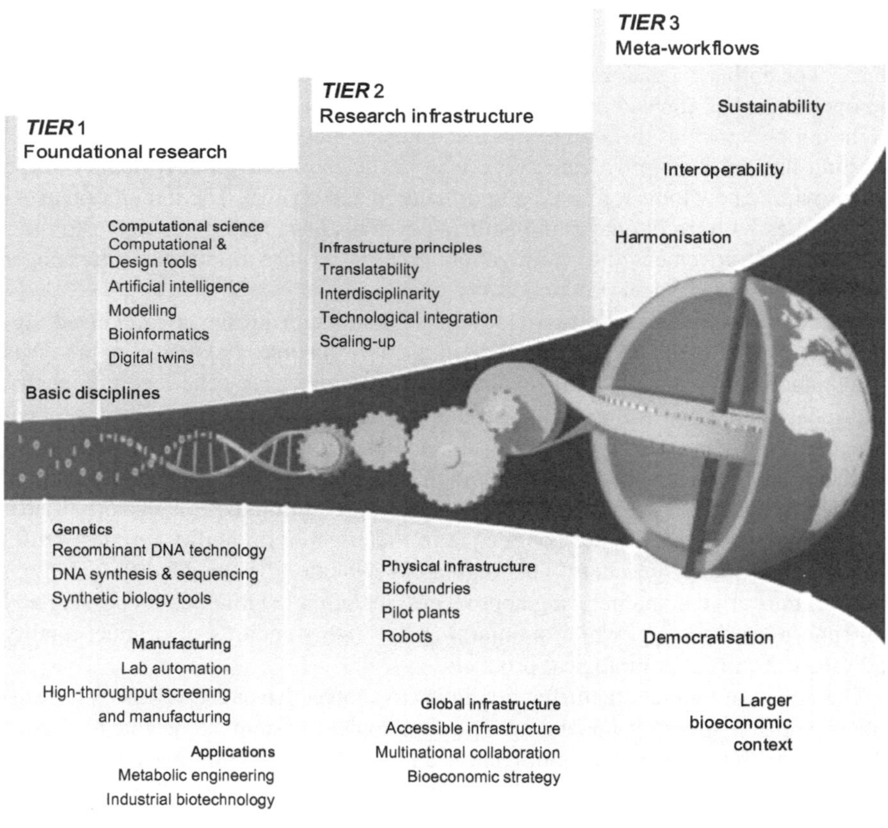

FIGURE 1.2 Integration of key principles in modern biomanufacturing systems.

numerous considerations, from cell line development and process optimization to facility design and regulatory compliance, creating a multifaceted approach to biological product manufacturing that requires expertise across multiple disciplines.

The historical development of biomanufacturing principles reflects the industry's journey from simple fermentation processes to today's highly controlled and monitored manufacturing systems. This evolution has been marked by significant advances in process understanding and control, enabled by improvements in analytical technologies and process monitoring capabilities. The early days of biomanufacturing, characterized by limited process understanding and rudimentary control strategies, have given way to sophisticated approaches that leverage advanced technologies and systematic methodologies to ensure product quality and process consistency. This transformation has been particularly evident in the biopharmaceutical sector, where the demands for product quality and regulatory compliance have driven continuous innovation in manufacturing approaches and technologies.

1.2.1 Process Development and Optimization Strategies

The development and optimization of biomanufacturing processes represents a critical aspect of biological product manufacturing, requiring careful consideration of numerous factors that can impact process performance and product quality. Process development typically begins with small-scale experiments designed to establish basic process parameters and understand critical quality attributes. This initial work provides the foundation for subsequent process optimization efforts, which aim to improve process efficiency while maintaining or enhancing product quality. The optimization process must consider multiple factors simultaneously, including cell line characteristics, media composition, process parameters, and equipment capabilities, creating a complex optimization landscape that requires sophisticated approaches to navigate effectively.

The application of systematic approaches to process development and optimization has become increasingly important as biomanufacturing processes become more complex. These approaches often incorporate design of experiments (DOE) methodologies and statistical analysis tools to understand the relationships between process parameters and product quality attributes. This systematic approach enables manufacturers to develop robust processes that can consistently produce high-quality products while maintaining operational efficiency. The integration of advanced analytical technologies and process monitoring systems has further enhanced these optimization efforts, providing manufacturers with unprecedented insight into their processes and enabling more effective process control strategies.

1.2.1.1 Quality Systems and Regulatory Considerations

The quality systems that support biomanufacturing operations represent a critical component of successful biological product manufacturing. These systems encompass all aspects of manufacturing operations, from raw material testing and process control to product testing and release (Table 1.1). The development and implementation of effective quality systems requires careful consideration of regulatory

TABLE 1.1
Critical Elements of Biomanufacturing Quality Systems

Quality Element	Implementation Strategy	Impact on Process	Regulatory Significance
Process controls	Real-time monitoring	Product consistency	GMP compliance
Documentation	Electronic systems	Data integrity	Regulatory review
Training	Competency programs	Operational excellence	Personnel qualification
Validation	Life cycle approach	Process robustness	Product quality

requirements, operational needs, and practical constraints, creating a framework that ensures consistent product quality while maintaining operational efficiency. The integration of quality systems with manufacturing operations has become increasingly important as processes have become more complex and regulatory requirements more stringent.

1.2.2 Facility Design and Operations Management

The design and operation of biomanufacturing facilities represents a significant challenge that requires careful consideration of numerous factors, from process requirements and containment needs to regulatory compliance and operational efficiency. Modern biomanufacturing facilities must incorporate sophisticated environmental control systems, advanced process equipment, and robust monitoring capabilities while maintaining the flexibility needed to adapt to changing process requirements and product demands. The integration of these various elements creates complex operational challenges that must be carefully managed to ensure successful manufacturing operations.

The implementation of effective facility management strategies has become increasingly important as biomanufacturing operations become more complex. These strategies must address numerous considerations, from equipment maintenance and calibration to personnel training and operational procedures. The development of comprehensive facility management programs requires careful consideration of operational needs, regulatory requirements, and resource constraints, creating a framework that ensures reliable facility operation while maintaining compliance with applicable regulations.

1.2.2.1 Technology Integration and Advanced Manufacturing Systems

The integration of advanced technologies in biomanufacturing operations has created new opportunities for process improvement and optimization while also introducing new challenges that must be carefully managed. Modern biomanufacturing facilities increasingly incorporate sophisticated automation systems, advanced process controls, and integrated data management capabilities, creating complex technological ecosystems that require careful management and maintenance. The successful implementation of these technologies requires careful consideration of numerous factors, from system validation and data integrity to personnel training and change management.

The emergence of artificial intelligence and machine learning applications in biomanufacturing represents a significant advancement in process control and optimization capabilities. These technologies enable the analysis of complex process data and the identification of subtle patterns and relationships that might not be apparent through traditional analysis methods. The integration of AI technologies with traditional manufacturing systems has created new opportunities for process improvement and optimization, pointing toward a future of increasingly sophisticated and efficient production systems.

1.2.3 Knowledge Management and Continuous Improvement

The management of knowledge and implementation of continuous improvement programs represent a critical aspect of successful biomanufacturing operations. These programs must address numerous considerations, from process understanding and optimization to personnel training and development. The implementation of effective knowledge management systems requires careful consideration of organizational needs, technological capabilities, and regulatory requirements, creating a framework that supports continuous learning and improvement while maintaining operational efficiency.

The development of comprehensive training programs and knowledge transfer systems has become increasingly important as biomanufacturing operations become more complex. These programs must address numerous considerations, from basic operational procedures to advanced process understanding and troubleshooting capabilities. The successful implementation of these programs requires careful consideration of learning objectives, training methodologies, and assessment approaches, creating a framework that ensures effective knowledge transfer while maintaining operational efficiency.

1.3 OVERVIEW OF BIOMANUFACTURING

The evolution of biomanufacturing represents a remarkable journey from traditional fermentation processes to today's sophisticated, automated production systems. This transformation has been driven by advances in multiple disciplines, including molecular biology, process engineering, and automation technology. The early stages of biomanufacturing were characterized by simple batch processes with limited process understanding and control capabilities. These rudimentary systems have evolved into highly sophisticated manufacturing operations that leverage advanced technologies and systematic approaches to ensure consistent product quality. The progression from traditional manufacturing methods to modern biomanufacturing systems reflects not only technological advancement but also our growing understanding of biological systems and their behavior in industrial-scale processes.

The current state of biomanufacturing represents the culmination of decades of technological advancement and process innovation. Modern biomanufacturing facilities incorporate sophisticated control systems, advanced analytical technologies, and robust quality management systems to ensure consistent product quality. These facilities must meet stringent regulatory requirements while maintaining operational efficiency and flexibility to accommodate different products and processes. The integration of advanced technologies has created new opportunities for process optimization and improvement while also introducing new challenges that must be carefully managed. The complexity of modern biomanufacturing operations requires a comprehensive understanding of biological systems, process engineering principles, and quality management approaches (Figure 1.3).

1.3.1 Cellular and Molecular Foundations

The cellular and molecular foundations of biomanufacturing processes represent critical elements that must be thoroughly understood to ensure successful manufacturing operations. The behavior of cells in industrial-scale processes can differ significantly from their behavior in laboratory conditions, requiring careful consideration of numerous factors that can impact cell growth and product expression. The optimization of cell culture conditions requires a deep understanding of cellular metabolism, growth kinetics, and protein expression patterns. This understanding must extend beyond basic biological principles to encompass the complex interactions between cells and their environment in industrial-scale processes.

The selection and development of production cell lines represents a critical aspect of biomanufacturing process development. The characteristics of production cell lines can significantly impact process performance and product quality, requiring careful consideration during cell line development and selection. The stability of production cell lines over extended culture periods must be carefully evaluated to ensure consistent product quality throughout the manufacturing process. The optimization of cell line characteristics requires consideration of numerous factors, from growth characteristics and protein expression levels to genetic stability and metabolic profiles (Table 1.2).

1.3.2 Process Engineering Fundamentals

The engineering principles that underpin biomanufacturing processes represent essential elements of successful manufacturing operations. These principles encompass numerous aspects of process design and control, from basic mass transfer and heat transfer considerations to complex control system design and implementation. The application of engineering principles in biomanufacturing requires careful consideration of the unique characteristics of biological systems and their behavior in industrial-scale processes. The integration of engineering principles with biological understanding creates a framework for process design and optimization that ensures consistent product quality while maintaining operational efficiency.

Traditional Biotechnology	Modern Biotechnology
It refers to the traditional techniques of using living organisms to yield new products or modify foods or other useful products for human use.	The diversity of microorganisms and development of genetics expanded the potential of traditional biotech, and ultimately led to the development of modern biotechnology.
Examples of early biotechnology include recombinant DNA techniques, tissue culture, and mutagenesis.	Examples of modern biotechnology include DNA profiling, genome analysis, transgenesis, DNA cloning, tissue engineering, etc.
Breeding animals and crops to make cheese, yoghurt, bread, beer and wine. Microorganisms are used to produce various products such as enzymes for use in laundry detergents.	Modern biotechnology techniques are used in many areas such as food, agriculture, forestry, healthcare, environment, minerals, and industrial processes.
It involves use of natural organisms to create or modify food or other useful products for human use.	It involves manipulation of genes and living tissues in a controlled environment.

FIGURE 1.3 Technological evolution in biomanufacturing: From traditional to modern systems.

TABLE 1.2
Critical Aspects of Cell Line Development and Maintenance

Aspect	Consideration Factors	Impact on Process	Quality Implications
Growth characteristics	Media requirements	Production efficiency	Process consistency
Expression stability	Genetic factors	Product quality	Manufacturing duration
Metabolic profile	Nutrient utilization	Process control	Product attributes
Cell line security	Storage conditions	Long-term stability	Risk management

The scale-up of biomanufacturing processes from laboratory scale to commercial production represents a significant engineering challenge that requires careful consideration of numerous factors. The behavior of biological systems can change significantly as process scale increases, requiring careful evaluation of scale-up parameters and their impact on process performance. The development of robust scale-up strategies requires consideration of numerous factors, from mixing and mass transfer characteristics to heat transfer and oxygen delivery capabilities. The successful scale-up of biomanufacturing processes requires a comprehensive understanding of both biological and engineering principles.

1.3.3 Equipment and Facility Infrastructure

Modern biomanufacturing facilities require sophisticated equipment and infrastructure systems to support successful manufacturing operations. The design and implementation of these systems must consider numerous factors, from basic utility requirements to advanced process control capabilities. The integration of various equipment and infrastructure systems creates complex operational requirements that must be carefully managed to ensure reliable facility operation. The maintenance and calibration of manufacturing equipment represents a critical aspect of facility operations that requires careful attention to ensure consistent process performance.

The design of clean rooms and controlled environments represents a critical aspect of biomanufacturing facility design. These systems must maintain appropriate environmental conditions while facilitating efficient manufacturing operations. The implementation of effective contamination control strategies requires careful consideration of numerous factors, from air handling and filtration systems to personnel and material flows. The validation of clean room operations requires comprehensive testing and monitoring programs to ensure consistent environmental conditions.

1.3.4 Analytical Technologies and Process Monitoring

The implementation of advanced analytical technologies and process monitoring systems represents a critical aspect of modern biomanufacturing operations. These systems provide essential information about process performance and product quality, enabling effective process control and optimization. The selection and implementation of appropriate analytical technologies requires careful consideration of numerous factors, from measurement capabilities and accuracy requirements to sampling strategies and data management approaches. The integration of analytical technologies with process control systems creates powerful tools for process monitoring and control.

The development of robust sampling and testing strategies represents an essential element of process monitoring programs. These strategies must consider numerous factors, from sample representativeness and stability to analytical method capabilities and turnaround time requirements. The implementation of effective sampling programs requires careful consideration of operational constraints and regulatory requirements while maintaining the ability to generate meaningful process information. The interpretation of analytical results requires careful consideration of measurement uncertainties and process variability.

1.3.5 Quality Management Systems Integration

The integration of quality management systems with manufacturing operations represents a fundamental aspect of successful biomanufacturing. These systems must ensure consistent product quality while maintaining operational efficiency and regulatory compliance. The development of comprehensive quality systems requires careful consideration of numerous factors, from documentation requirements and training programs to change control and deviation management processes. The implementation of effective quality systems requires engagement from all levels of the organization and a commitment to continuous improvement.

The management of process deviations and investigations represents a critical aspect of quality systems operation. These activities must identify root causes of process deviations while implementing effective corrective and preventive actions. The investigation of process deviations requires careful evaluation of available data and potential contributing factors while maintaining compliance with regulatory requirements. The implementation of effective corrective actions requires careful consideration of process impacts and validation requirements.

1.3.6 Regulatory Compliance and Documentation

The maintenance of regulatory compliance represents a fundamental requirement for biomanufacturing operations. These requirements encompass numerous aspects of manufacturing operations, from facility design and equipment qualification to process validation and product testing. The implementation of effective compliance programs requires careful consideration of regulatory requirements and their practical implementation in manufacturing operations. The development of comprehensive documentation systems represents an essential element of regulatory compliance programs.

The management of manufacturing documentation requires careful attention to numerous factors, from document control and retention requirements to review and approval processes. The implementation of effective documentation systems requires consideration of operational needs and regulatory requirements while maintaining the ability to demonstrate process control and product quality. The review and approval of manufacturing documentation requires careful evaluation of content completeness and accuracy while maintaining compliance with regulatory requirements.

1.4 THE QUALITY BY DESIGN APPROACH

The quality by design (QbD) approach represents a paradigm shift in biomanufacturing, moving away from traditional quality-by-testing methods toward a more systematic and science-based approach to product development and manufacturing. This comprehensive methodology emphasizes building quality into products from the earliest stages of development, rather than attempting to test quality into final products. The implementation of QbD principles in biomanufacturing has revolutionized how companies approach process development, validation, and control, creating a framework that ensures consistent product quality while maintaining operational efficiency. This transformation has been particularly significant in the biopharmaceutical industry, where the complexity of biological products and manufacturing processes creates unique challenges for quality assurance and control.

The historical context of QbD implementation in biomanufacturing provides valuable insights into its evolution and impact on the industry. Initially introduced through regulatory guidance documents, QbD has grown from a theoretical concept into a practical methodology that guides process development and manufacturing decisions across the industry. This evolution has been driven by increasing recognition of the benefits of systematic approaches to quality management, including improved process understanding, reduced manufacturing variability, and more efficient regulatory compliance. The adoption of QbD principles has led to significant improvements in manufacturing efficiency and product quality while also facilitating more effective communication between manufacturers and regulatory agencies.

The foundation of QbD rests upon a deep understanding of product and process characteristics, achieved through systematic application of scientific principles and risk management approaches. This understanding begins with clear definition of product quality attributes and extends through all aspects of process development and control. The systematic nature of QbD ensures that critical quality attributes are identified early in development and that appropriate control strategies are implemented to ensure consistent product quality. This comprehensive approach requires significant investment in process understanding and control capabilities but typically results in more robust and efficient manufacturing processes.

1.4.1 Risk Assessment and Management in QbD Implementation

Risk assessment and management form crucial components of the QbD approach, providing systematic methods for identifying and addressing potential quality issues throughout the product life cycle. The implementation of effective risk management strategies requires careful consideration of numerous factors, from process variability and control capabilities to regulatory requirements and operational constraints. These assessments typically begin during early process development and continue throughout the product life cycle, creating a dynamic framework for identifying and managing quality risks. The integration of risk management principles with process development and control strategies ensures that critical quality attributes are maintained while operational efficiency is optimized.

The development of comprehensive risk assessment programs requires careful consideration of multiple risk factors and their potential interactions. These programs must address both technical and operational risks, from process variability and equipment reliability to personnel training and supplier quality. The implementation of effective risk management strategies requires systematic evaluation of risk factors and development of appropriate mitigation strategies. This systematic approach enables manufacturers to identify potential quality issues before they impact product quality, allowing for proactive implementation of control strategies and corrective actions (Figure 1.4).

1.4.2 Process Understanding and Control Strategy Development

The development of comprehensive process understanding represents a fundamental aspect of QbD implementation, requiring systematic investigation of process parameters and their relationships to product quality attributes. This understanding begins with careful characterization of critical process parameters and extends through development of appropriate control strategies. The systematic nature of this approach ensures that process variables are well understood and that appropriate control strategies are implemented to maintain product quality. The integration of process understanding with control strategy development creates a robust framework for ensuring consistent product quality while maintaining operational efficiency.

The implementation of effective control strategies requires careful consideration of numerous factors, from process capability and variability to monitoring requirements and control limits. These strategies must address all aspects of the manufacturing process, from raw material quality to final product testing. The development of comprehensive control strategies requires systematic evaluation of process parameters and their relationships to product quality attributes. This systematic approach enables manufacturers to implement effective control strategies that ensure consistent product quality while maintaining operational efficiency (Table 1.3).

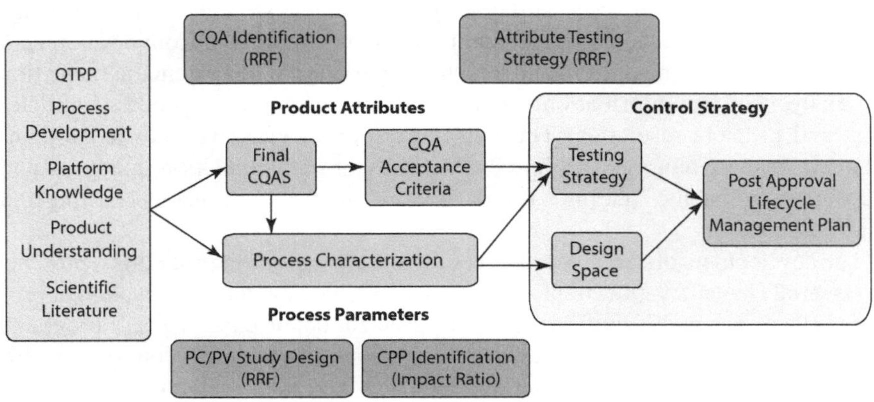

FIGURE 1.4 QbD principles in biomanufacturing process development.

TABLE 1.3
Critical Elements of QbD Process Control Strategy

Control Element	Implementation Approach	Quality Impact	Monitoring Requirements
Process parameters	Design space definition	Product attributes	Real-time monitoring
Material controls	Supplier quality systems	Raw material quality	Testing programs
Environmental controls	Facility systems	Process consistency	Continuous monitoring
Personnel controls	Training programs	Operational excellence	Competency assessment

1.4.3 Design Space Development and Validation

The concept of design space represents a key element of QbD implementation, providing a scientific framework for understanding and controlling process parameters that impact product quality. The development of appropriate design spaces requires careful consideration of numerous factors, from process capability and variability to regulatory requirements and operational constraints. These design spaces typically incorporate multiple process parameters and their interactions, creating a multidimensional framework for process control and optimization. The validation of design spaces requires systematic evaluation of process performance across the defined operating ranges, ensuring that product quality is maintained throughout the design space.

1.4.4 Knowledge Management and Continuous Improvement

The management of knowledge generated through QbD implementation represents a critical aspect of successful quality management systems. This knowledge encompasses all aspects of product and process understanding, from development studies through commercial manufacturing experience. The systematic collection and analysis of this information creates a valuable resource for process improvement and problem resolution. The implementation of effective knowledge management systems requires careful consideration of data collection, analysis, and sharing requirements, creating a framework that supports continuous learning and improvement.

The development of comprehensive continuous improvement programs represents an essential element of QbD implementation, requiring systematic evaluation of process performance and identification of improvement opportunities. These programs must address all aspects of manufacturing operations, from process control and optimization to personnel training and development. The implementation of effective improvement programs requires careful consideration of improvement objectives, implementation strategies, and success metrics, creating a framework that supports ongoing process optimization and quality enhancement.

1.4.5 Regulatory Considerations and Compliance Strategy

The regulatory landscape surrounding QbD implementation continues to evolve, requiring manufacturers to maintain the current understanding of regulatory expectations and requirements. The development of appropriate compliance strategies requires careful consideration of numerous factors, from regulatory guidance documents to specific product requirements. These strategies must address all aspects of QbD implementation, from development studies through commercial manufacturing. The integration of regulatory considerations with QbD implementation ensures that quality systems meet both regulatory requirements and operational needs.

The implementation of effective regulatory compliance programs requires systematic evaluation of regulatory requirements and development of appropriate compliance strategies. These programs must address numerous considerations, from documentation requirements to change management procedures. The development of comprehensive compliance strategies requires careful consideration of regulatory expectations and operational capabilities, creating a framework that ensures compliance while maintaining operational efficiency. This systematic approach enables manufacturers to meet regulatory requirements while maintaining flexibility for process improvement and optimization.

1.5 TECHNICAL CONSIDERATIONS IN BIOMANUFACTURING

1.5.1 Process Control and Monitoring Systems

The implementation of sophisticated process control and monitoring systems represents one of the most critical technical aspects of modern biomanufacturing operations. These systems form the backbone of successful biological product manufacturing, enabling precise control over critical process parameters while providing real-time insights into process performance and product quality. The complexity of biological systems necessitates a multifaceted approach to process control, incorporating various sensors, monitoring devices, and control algorithms that work in concert to maintain optimal processing conditions. Advanced monitoring systems continuously track numerous parameters, including temperature, pH, dissolved oxygen, nutrient concentrations, and metabolic indicators, generating vast amounts of data that must be effectively managed and analyzed to ensure process consistency and product quality.

The evolution of process control systems in biomanufacturing has been marked by significant advances in sensor technology and data analysis capabilities. Modern facilities employ a wide array of in-line, on-line, and at-line monitoring systems that provide continuous feedback on process performance and product quality attributes. These advanced monitoring capabilities have enabled the development of more sophisticated control strategies, moving beyond simple feedback control to incorporate predictive and adaptive control mechanisms that can anticipate and respond to process variations before they impact product quality. The integration of these various monitoring and control systems creates a complex technological ecosystem that requires careful management and maintenance to ensure reliable operation (Figure 1.5).

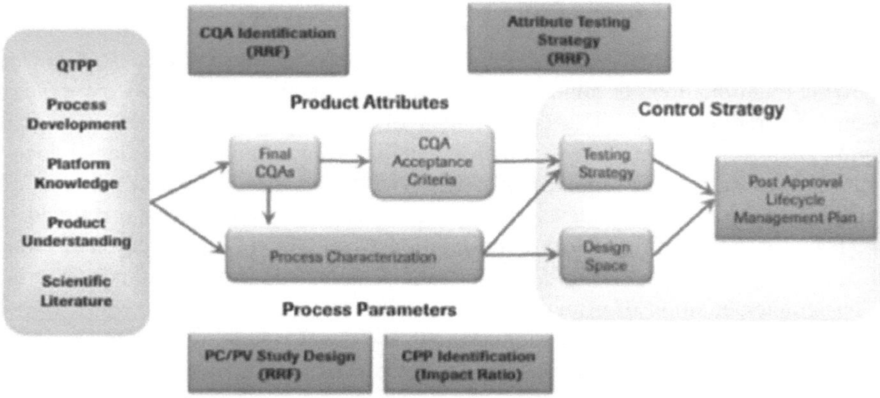

FIGURE 1.5 Integrated process control and monitoring architecture.

1.5.2 Equipment Design and Validation Strategies

The design and validation of biomanufacturing equipment presents unique challenges that must be carefully addressed to ensure successful operation and regulatory compliance. Equipment design must consider numerous factors, from basic operational requirements and maintenance needs to more complex considerations such as cleanability, sterilization capabilities, and material compatibility. The development of effective validation strategies requires a thorough understanding of equipment capabilities and limitations as well as careful consideration of regulatory requirements and operational constraints. Modern biomanufacturing equipment often incorporates sophisticated control systems and monitoring capabilities that must be properly validated to ensure reliable operation and data integrity.

The implementation of comprehensive equipment qualification programs represents a critical aspect of successful biomanufacturing operations. These programs must address all aspects of equipment performance and reliability, from basic installation and operational qualification to more complex performance qualification activities. The development of effective qualification protocols requires careful consideration of equipment specifications, operational requirements, and regulatory expectations, creating a framework that ensures equipment reliability while maintaining compliance with applicable regulations. The integration of equipment qualification activities with broader validation programs creates additional complexity that must be carefully managed to ensure successful outcomes.

1.5.3 Material Management and Supply Chain Considerations

The management of materials in biomanufacturing operations represents a critical technical consideration that impacts both process performance and product quality. Material management systems must address numerous requirements, from proper storage and handling procedures to detailed tracking and documentation capabilities.

TABLE 1.4
Critical Material Management Parameters in Biomanufacturing

Material Category	Storage Requirements	Quality Controls	Supply Chain Considerations
Raw materials	Temperature controlled	Identity testing	Lead time management
Cell banks	Cryogenic storage	Viability testing	Backup storage sites
Process solutions	Controlled environment	Sterility testing	Just-in-time delivery
Consumables	Ambient conditions	Compatibility testing	Inventory management

The implementation of effective material management strategies requires careful consideration of material characteristics, storage requirements, and usage patterns, creating a framework that ensures material quality while maintaining operational efficiency. The complexity of biological products and processes often necessitates specialized material handling and storage capabilities that must be carefully designed and maintained (Table 1.4).

1.5.4 Analytical Methods and Testing Strategies

The development and implementation of robust analytical methods represents a fundamental technical requirement in biomanufacturing operations. These methods must provide accurate and reliable measurements of numerous product quality attributes while maintaining the sensitivity and specificity needed to detect potential impurities or contaminants. The complexity of biological products often necessitates the use of multiple analytical techniques, creating additional challenges in method development and validation. The integration of analytical testing with manufacturing operations requires careful consideration of sampling strategies, testing schedules, and data management capabilities.

The validation of analytical methods represents another critical technical consideration in biomanufacturing operations. Method validation programs must demonstrate the reliability and reproducibility of analytical results while establishing appropriate acceptance criteria for various quality attributes. The development of comprehensive validation protocols requires careful consideration of method capabilities, regulatory requirements, and operational constraints, creating a framework that ensures reliable analytical results while maintaining compliance with applicable regulations. The integration of analytical method validation with broader quality control programs creates additional complexity that must be carefully managed (Figure 1.6).

1.5.5 Contamination Control and Environmental Monitoring

The implementation of effective contamination control strategies represents a critical technical requirement in biomanufacturing operations. These strategies must address numerous potential contamination sources, from raw materials and process

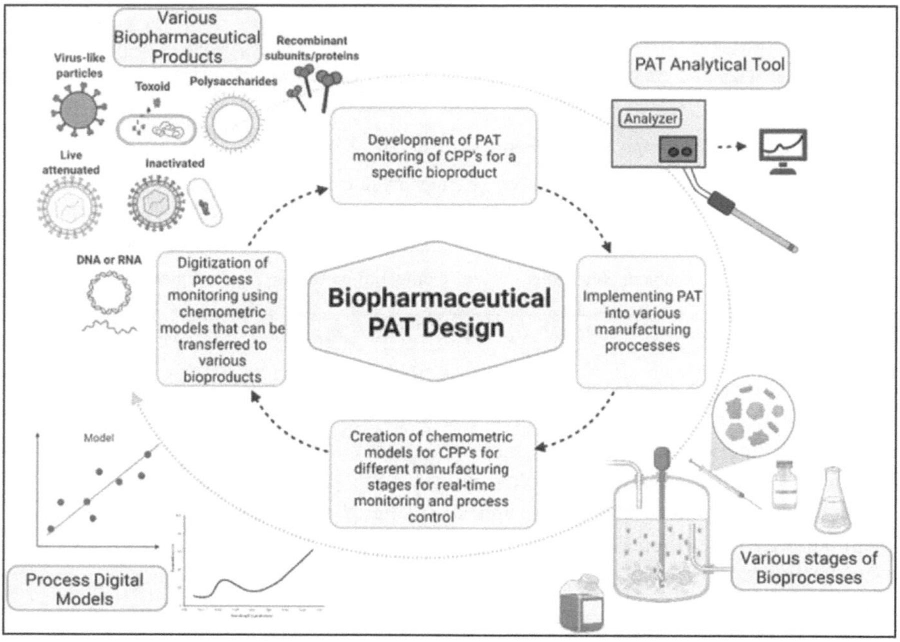

FIGURE 1.6 Integrated analytical testing strategy for biomanufacturing process control.

equipment to personnel and environmental factors. The development of comprehensive contamination control programs requires careful consideration of facility design, operational procedures, and monitoring capabilities, creating a framework that ensures product quality while maintaining operational efficiency. The complexity of biological processes often necessitates specialized containment and monitoring capabilities that must be carefully designed and maintained.

Environmental monitoring programs represent another critical technical consideration in biomanufacturing operations. These programs must provide comprehensive coverage of manufacturing areas while maintaining the sensitivity needed to detect potential contamination events. The development of effective monitoring strategies requires careful consideration of sampling locations, monitoring frequencies, and alert/action limits, creating a framework that ensures early detection of potential problems while maintaining operational efficiency. The integration of environmental monitoring with broader contamination control programs creates additional complexity that must be carefully managed.

1.5.6 Data Management and Information Systems

The management of manufacturing data represents a critical technical consideration that impacts all aspects of biomanufacturing operations. Modern facilities generate vast amounts of process data that must be effectively collected, analyzed, and stored to support process control and regulatory compliance requirements. The implementation of robust data management systems requires careful consideration of system architecture, data integrity controls, and security requirements, creating a framework that ensures reliable data management while maintaining operational efficiency. The complexity of biomanufacturing operations often necessitates specialized data management capabilities that must be carefully designed and maintained.

The development of comprehensive information management strategies represents another critical technical consideration in biomanufacturing operations. These strategies must address numerous requirements, from basic data collection and storage to more complex analysis and reporting capabilities. The implementation of effective information management systems requires careful consideration of user needs, regulatory requirements, and technological capabilities, creating a framework that supports decision-making while maintaining compliance with applicable regulations. The integration of information management systems with manufacturing operations creates additional complexity that must be carefully managed to ensure successful outcomes.

1.6 PHASES AND SCALE-UP IN BIOMANUFACTURING

1.6.1 Fundamentals of Biomanufacturing Scale-Up

The process of scaling up biomanufacturing operations represents one of the most critical and challenging aspects of biological product development. This complex transition from laboratory-scale processes to commercial production encompasses numerous technical, operational, and regulatory considerations that must be carefully balanced to ensure successful outcomes. The fundamental principles of scale-up in biomanufacturing extend far beyond simple volumetric increases, requiring deep understanding of how process parameters and cellular responses change across different scales of operation. Success in scale-up operations demands careful consideration of numerous factors, including oxygen transfer rates, mixing patterns, heat transfer characteristics, and cellular metabolism, all of which can significantly impact process performance and product quality.

The historical evolution of scale-up methodologies in biomanufacturing reflects the industry's growing understanding of the complex relationships between process scale and product quality. Early approaches to scale-up often relied heavily on empirical observations and trial-and-error methodologies, leading to frequent challenges in maintaining product quality and process consistency across different scales of operation. Modern scale-up strategies incorporate sophisticated modeling techniques and systematic approaches to process characterization, enabling more predictable and successful transitions between different scales of operation. This evolution has been particularly significant in the biopharmaceutical sector, where the demands for product quality and process consistency have driven continuous innovation in scale-up methodologies.

1.6.2 Process Development and Characterization across Scales

The development and characterization of biomanufacturing processes across different scales of operation requires comprehensive understanding of how process parameters and cellular responses change as scale increases. This understanding begins with detailed characterization of laboratory-scale processes, establishing baseline performance metrics and identifying critical process parameters that must be carefully controlled during scale-up operations. The translation of these parameters to larger scales requires careful consideration of numerous factors, including equipment capabilities, operational constraints, and regulatory requirements, creating a complex optimization challenge that must be carefully managed to ensure successful outcomes.

Process characterization activities must address numerous aspects of manufacturing operations, from basic cellular metabolism and growth characteristics to complex interactions between process parameters and product quality attributes. This comprehensive approach to process characterization enables manufacturers to develop robust scale-up strategies that can reliably produce high-quality products across different scales of operation. The integration of advanced analytical technologies and process monitoring systems has further enhanced these characterization efforts, providing manufacturers with unprecedented insight into their processes and enabling more effective scale-up strategies.

1.6.3 Equipment and Facility Considerations in Scale-Up

The selection and implementation of appropriate equipment and facility systems represents a critical aspect of successful scale-up operations in biomanufacturing. Equipment considerations must address numerous factors, from basic operational capabilities to advanced control features and monitoring systems. The integration of these various elements creates complex technical challenges that must be carefully managed to ensure successful scale-up operations. Modern biomanufacturing facilities must incorporate sophisticated environmental control systems, advanced process equipment, and robust monitoring capabilities while maintaining the flexibility needed to support effective scale-up operations (Table 1.5).

TABLE 1.5
Critical Equipment Parameters across Manufacturing Scales

Scale Level	Equipment Type	Key Parameters	Scale-Up Challenges
Laboratory	Bench reactors	Basic control	Process validation
Pilot	Mid-size systems	Enhanced monitoring	Parameter translation
Commercial	Production units	Integrated control	System integration
Manufacturing	Industrial scale	Advanced automation	Process consistency

1.6.4 Operational Challenges and Risk Management

The operational challenges associated with scale-up operations in biomanufacturing require careful consideration of numerous factors that can impact process performance and product quality. These challenges extend across multiple aspects of manufacturing operations, from basic process control and monitoring to complex troubleshooting and risk management activities. The development of effective risk management strategies requires careful consideration of potential failure modes and their impact on process performance and product quality, creating a framework that ensures reliable manufacturing operations while maintaining compliance with applicable regulations.

Risk assessment and mitigation strategies play a crucial role in successful scale-up operations, requiring careful evaluation of potential risks and development of appropriate control strategies. These assessments must consider numerous factors, from basic process parameters and equipment capabilities to complex interactions between different aspects of manufacturing operations. The implementation of effective risk management programs requires careful consideration of operational needs, regulatory requirements, and resource constraints, creating a framework that supports successful scale-up operations while maintaining appropriate levels of control and oversight (Figure 1.7).

1.6.5 Technology Transfer and Process Validation

The transfer of manufacturing processes between different scales and facilities represents a critical aspect of successful scale-up operations in biomanufacturing. Technology transfer activities must address numerous considerations, from basic process parameters and equipment capabilities to complex validation requirements

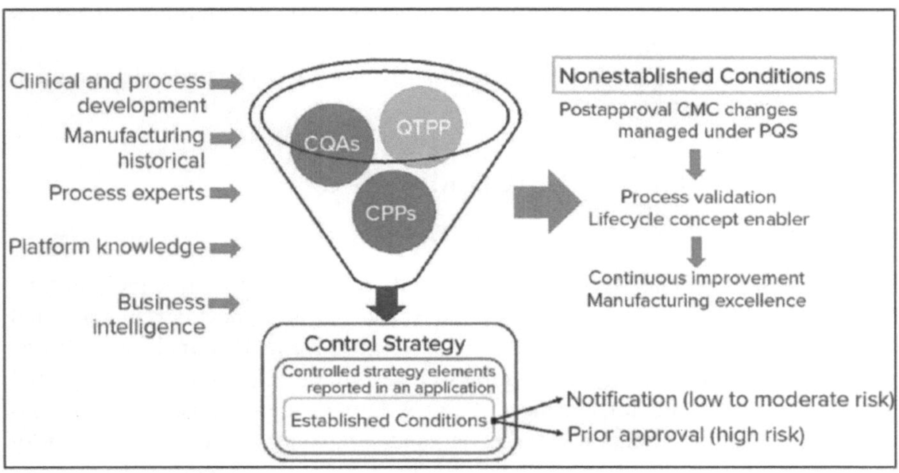

FIGURE 1.7 Risk management framework for biomanufacturing scale-up operations.

and regulatory considerations. The development of comprehensive technology transfer protocols requires careful consideration of numerous factors, ensuring successful translation of manufacturing processes while maintaining product quality and regulatory compliance.

Process validation activities play a crucial role in demonstrating the reliability and consistency of scaled-up manufacturing processes. These activities must address numerous aspects of manufacturing operations, from basic process parameters and control strategies to complex monitoring systems and data management capabilities. The implementation of effective validation programs requires careful consideration of regulatory requirements, operational needs, and resource constraints, creating a framework that ensures reliable manufacturing operations while maintaining compliance with applicable regulations.

1.6.6 Continuous Improvement and Process Optimization

The implementation of continuous improvement programs represents an essential aspect of successful scale-up operations in biomanufacturing. These programs must address numerous considerations, from basic process optimization and efficiency improvements to complex technological innovations and operational enhancements. The development of effective improvement strategies requires careful consideration of numerous factors, from process understanding and optimization opportunities to resource constraints and regulatory requirements, creating a framework that supports ongoing process enhancement while maintaining operational efficiency.

Knowledge management and training programs play a crucial role in supporting successful scale-up operations, requiring careful consideration of learning objectives, training methodologies, and assessment approaches. The implementation of effective knowledge transfer systems requires careful consideration of organizational needs, technological capabilities, and regulatory requirements, creating a framework that supports continuous learning and improvement while maintaining operational efficiency. These programs must address numerous aspects of manufacturing operations, from basic operational procedures to advanced process understanding and troubleshooting capabilities.

1.7 LIFE CYCLE OF MANUFACTURING

1.7.1 Strategic Planning and Initial Development

The life cycle of biomanufacturing begins with comprehensive strategic planning that establishes the foundational framework for successful product development and commercialization. This critical phase involves extensive market analysis, resource assessment, and technological evaluation to determine the feasibility and potential success of proposed manufacturing initiatives. Organizations must carefully consider multiple factors, including market demand, competitive landscape, regulatory requirements, and available technologies, to develop effective manufacturing strategies that align with their overall business objectives. The strategic planning phase also encompasses the evaluation of existing manufacturing capabilities, identification of potential gaps, and development of plans to address these gaps through technology acquisition, facility upgrades, or strategic partnerships.

During the initial development phase, organizations must establish robust frameworks for product and process development that will support the entire manufacturing life cycle. This includes the development of comprehensive documentation systems, quality management frameworks, and risk assessment methodologies that will guide all subsequent manufacturing activities. The importance of this foundational work cannot be overstated, as decisions made during this phase will have long-lasting implications for manufacturing success. Organizations must carefully balance multiple considerations, including resource allocation, timeline requirements, and quality expectations, to establish effective development frameworks that support their manufacturing objectives (Figure 1.8).

1.7.2 Process Development and Technology Transfer

The process development phase represents a critical junction in the manufacturing life cycle, where theoretical concepts and laboratory-scale processes are transformed into practical manufacturing procedures. This phase requires extensive experimental work, data analysis, and process optimization to establish robust manufacturing processes that can consistently produce high-quality products. Organizations must carefully consider numerous factors, including scalability, reproducibility, and cost-effectiveness, when developing manufacturing processes. The integration of quality

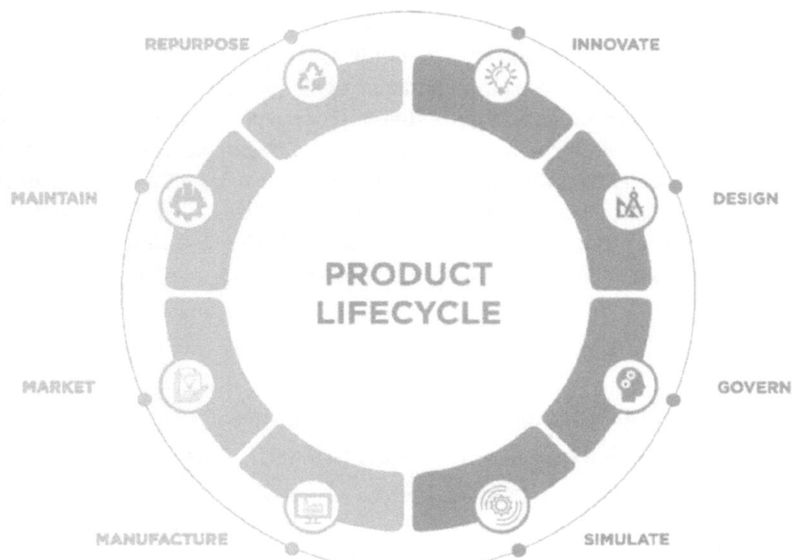

FIGURE 1.8 Comprehensive manufacturing life cycle framework.

by design principles during process development helps ensure that resulting processes are robust and capable of consistently meeting product quality requirements.

Technology transfer activities represent another crucial aspect of the manufacturing life cycle, requiring careful coordination between development and manufacturing teams to ensure successful implementation of established processes. This phase involves extensive documentation, training, and validation activities to ensure that manufacturing processes can be successfully executed in their intended production environment. Organizations must carefully manage the transfer of knowledge, skills, and technologies between different groups and facilities, ensuring that critical process understanding is effectively communicated and maintained throughout the transfer process.

1.7.3 Validation and Qualification Programs

The implementation of comprehensive validation and qualification programs represents a fundamental requirement for successful manufacturing operations. These programs must address all aspects of manufacturing operations, from facility and equipment qualification to process validation and cleaning verification. The development of effective validation strategies requires careful consideration of regulatory requirements, operational needs, and resource constraints. Organizations must establish systematic approaches to validation that ensure compliance while maintaining operational efficiency (Table 1.6).

1.7.4 Commercial Manufacturing Operations

The execution of commercial manufacturing operations represents the culmination of extensive development and validation efforts, requiring careful coordination of multiple activities to ensure successful product manufacturing. This phase involves the implementation of established procedures, monitoring of process performance, and management of deviations to maintain consistent product quality. Organizations must carefully balance production demands with quality requirements, ensuring that manufacturing operations remain compliant while meeting business objectives. The management of commercial operations requires extensive coordination between different functional groups, from production and quality control to maintenance and engineering.

TABLE 1.6
Critical Elements of Manufacturing Validation Programs

Validation Element	Program Components	Success Criteria	Implementation Strategy
Process validation	Performance qualification	Process consistency	Risk-based approach
Cleaning validation	Residue limits	Product safety	Scientific rationale
Computer systems	Software validation	Data integrity	GAMP guidelines
Analytical methods	Method transfer	**Result accuracy**	Method life cycle

The integration of modern manufacturing execution systems has become increasingly important in managing commercial operations. These systems provide real-time monitoring capabilities, automated data collection, and integrated quality management functions that support efficient manufacturing operations. Organizations must carefully implement and maintain these systems to ensure reliable operation while maintaining compliance with applicable regulations. The successful management of commercial operations requires continuous monitoring and adjustment of manufacturing activities to optimize performance and maintain product quality.

1.7.5 Continuous Improvement and Performance Optimization

The implementation of continuous improvement programs represents a critical aspect of the manufacturing life cycle, requiring systematic approaches to identify and implement process improvements. These programs must address all aspects of manufacturing operations, from process performance and product quality to operational efficiency and cost-effectiveness. Organizations must establish effective mechanisms for gathering and analyzing performance data, identifying improvement opportunities, and implementing corrective actions. The development of comprehensive improvement programs requires careful consideration of multiple factors, including resource availability, regulatory requirements, and business objectives.

The application of modern analytical tools and methodologies has enhanced the effectiveness of continuous improvement efforts. These tools enable more sophisticated analysis of process data and identification of improvement opportunities that might not be apparent through traditional analysis methods. Organizations must carefully implement these tools and train personnel in their effective use to maximize the benefits of continuous improvement programs. The successful execution of improvement initiatives requires careful coordination between different functional groups and effective communication of program objectives and results (Figure 1.9).

1.7.6 Product Life Cycle Management and End-of-Life Considerations

The management of product life cycles represents a complex challenge that requires careful consideration of multiple factors, from market demands and competitive pressures to regulatory requirements and manufacturing capabilities. Organizations must establish effective systems for monitoring product performance, managing change requests, and implementing improvements throughout the product life cycle. The development of comprehensive life cycle management programs requires careful consideration of multiple stakeholders' needs and expectations, from customers and regulatory authorities to internal manufacturing and quality teams.

End-of-life considerations represent another critical aspect of the manufacturing life cycle that must be carefully managed to ensure appropriate transition of products and processes. This includes the development of effective strategies for managing product discontinuation, equipment decommissioning, and knowledge retention. Organizations must carefully plan and execute these activities to minimize disruption to ongoing operations while maintaining compliance with applicable

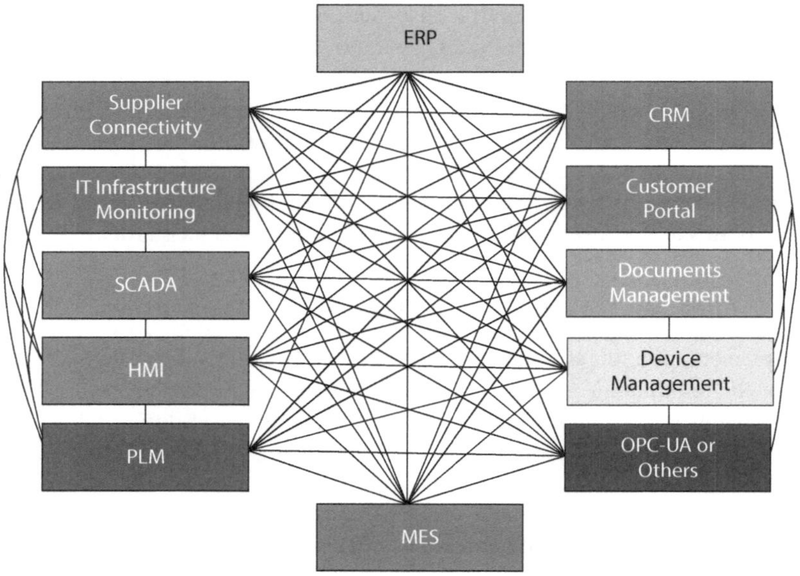

FIGURE 1.9 Integration of commercial manufacturing systems.

regulations. The successful management of end-of-life activities requires careful coordination between different functional groups and effective communication with all stakeholders.

1.7.7 Knowledge Management and Documentation Systems

The establishment of effective knowledge management systems represents a fundamental requirement for successful manufacturing operations. These systems must capture and maintain critical process knowledge, operational experience, and quality data throughout the manufacturing life cycle. Organizations must implement robust documentation systems that ensure appropriate recording and retention of manufacturing information while facilitating efficient access and use of this information. The development of effective knowledge management programs requires careful consideration of user needs, system capabilities, and regulatory requirements.

The integration of electronic systems has significantly enhanced the capabilities of knowledge management programs. These systems provide improved data accessibility, automated workflow management, and enhanced search capabilities that support efficient information management. Organizations must carefully implement and maintain these systems to ensure reliable operation while maintaining compliance with applicable regulations. The successful management of manufacturing knowledge requires continuous monitoring and updating of information to maintain its relevance and usefulness throughout the product life cycle.

1.8 INTEGRATION OF ARTIFICIAL INTELLIGENCE IN BIOMANUFACTURING

1.8.1 Evolution of AI Applications in Biomanufacturing

The integration of artificial intelligence in biomanufacturing represents a transformative advancement in how biological products are developed, manufactured, and controlled. This evolution has been driven by significant improvements in computational capabilities, data analytics, and machine learning algorithms, combined with the increasing complexity of biomanufacturing processes and the growing demand for process optimization and control. The journey of AI implementation in biomanufacturing began with simple statistical process control applications and has evolved into sophisticated systems capable of real-time process optimization and predictive maintenance. This transformation has been particularly impactful in the biopharmaceutical sector, where the complexity of biological processes and the stringent requirements for product quality create unique challenges that AI technologies are uniquely suited to address.

The historical progression of AI implementation in biomanufacturing provides valuable insights into both the potential and limitations of these technologies. Early applications focused primarily on data analysis and pattern recognition, using basic statistical algorithms to identify trends and relationships in process data. As computational capabilities improved and more sophisticated algorithms became available, the scope of AI applications expanded to include predictive modeling, process optimization, and real-time control systems. This evolution has been marked by increasing integration of AI technologies with traditional process control systems, creating hybrid approaches that leverage the strengths of both human expertise and artificial intelligence (Figure 1.10).

1.8.2 Machine Learning Applications in Process Development

The application of machine learning algorithms in process development has revolutionized how biomanufacturing processes are designed and optimized. These technologies enable rapid analysis of complex datasets, identification of critical process parameters, and development of predictive models that guide process development decisions. The implementation of machine learning approaches in process development requires careful consideration of data quality, algorithm selection, and validation requirements. The systematic application of these technologies has led to significant improvements in process understanding and development efficiency while also providing new insights into process behavior and control requirements.

Machine learning algorithms have proven particularly valuable in analyzing the complex relationships between process parameters and product quality attributes. These applications leverage advanced analytical capabilities to identify patterns and relationships that might not be apparent through traditional analysis methods. The integration of machine learning with process development activities has created new opportunities for process optimization and improvement while also providing

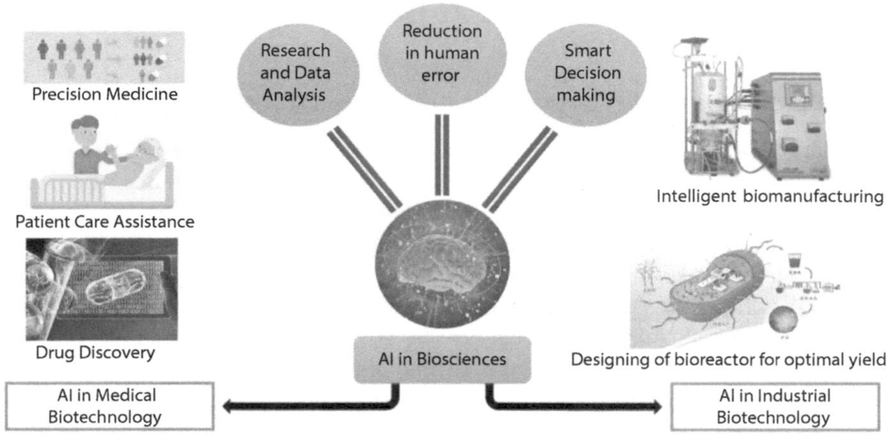

FIGURE 1.10 Evolution and integration of AI technologies in modern biomanufacturing.

valuable insights into process behavior and control requirements. This systematic approach to process development has led to more robust and efficient manufacturing processes.

1.8.3 Real-Time Process Control and Optimization

The implementation of AI-driven process control systems represents a significant advancement in biomanufacturing operations, enabling real-time optimization and control of complex biological processes. These systems leverage advanced algorithms and real-time data analysis capabilities to maintain optimal process conditions and ensure consistent product quality. The development of effective AI-driven control strategies requires careful consideration of process dynamics, control requirements, and validation needs. The integration of AI technologies with traditional control systems has created new opportunities for process improvement and optimization while also providing enhanced capabilities for process monitoring and control (Table 1.7).

TABLE 1.7
AI Applications in Process Control and Optimization

Application Area	AI Technology	Implementation Benefits	Operational Impact
Process control	Neural networks	Real-time optimization	Improved consistency
Quality prediction	Deep learning	Predictive analytics	Reduced variability
Fault detection	Pattern recognition	**Early warning systems**	Increased reliability
Process optimization	**Reinforcement learning**	Adaptive control	Enhanced efficiency

1.8.4 Predictive Maintenance and Equipment Management

The application of AI technologies in equipment maintenance and management has transformed how biomanufacturing facilities approach equipment reliability and maintenance planning. These systems leverage advanced analytics and machine learning algorithms to predict equipment failures and optimize maintenance schedules. The implementation of predictive maintenance programs requires careful consideration of equipment characteristics, failure modes, and maintenance requirements. The systematic application of AI technologies in equipment management has led to significant improvements in equipment reliability and maintenance efficiency while also reducing the risk of unexpected equipment failures.

1.8.5 Quality Assurance and Regulatory Compliance

The integration of AI technologies in quality assurance and regulatory compliance represents a significant advancement in how biomanufacturing companies approach quality management and regulatory requirements. These applications leverage advanced analytics and machine learning capabilities to enhance quality control processes and ensure regulatory compliance. The implementation of AI-driven quality systems requires careful consideration of regulatory requirements, data integrity needs, and validation approaches. The systematic application of AI technologies in quality management has led to significant improvements in quality assurance efficiency and effectiveness while also providing enhanced capabilities for regulatory compliance.

The development of comprehensive validation approaches for AI systems represents a critical aspect of successful implementation. These approaches must address numerous considerations, from algorithm validation and data integrity to system reliability and regulatory compliance. The implementation of effective validation strategies requires careful consideration of regulatory requirements, operational needs, and technical capabilities. This systematic approach ensures that AI systems meet both regulatory requirements and operational needs while maintaining the flexibility needed for continuous improvement and optimization.

1.8.6 Data Management and Integration Systems

The management and integration of data generated through AI applications represents a crucial aspect of successful implementation. These systems must handle vast amounts of process data while maintaining data integrity and accessibility. The development of effective data management strategies requires careful consideration of data collection, storage, and analysis requirements. The implementation of comprehensive data management systems ensures that AI applications have access to high-quality data while maintaining regulatory compliance and operational efficiency.

The integration of AI systems with existing manufacturing execution systems (MESs) and enterprise resource planning (ERP) systems represents a significant challenge that requires careful consideration of system architecture and data flow requirements. These integration efforts must address numerous considerations, from data format and transfer protocols to system security and validation requirements. The development of effective integration strategies ensures that AI applications can leverage existing data sources while maintaining system integrity and regulatory compliance.

1.9 CONCLUSION

The comprehensive exploration of biomanufacturing principles and practices presented in this chapter reveals the intricate complexity and dynamic nature of modern biological product manufacturing. The evolution of biomanufacturing from traditional processes to today's sophisticated, technology-driven operations represents a remarkable achievement in the integration of biological science, engineering principles, and advanced manufacturing technologies. This transformation has been driven by multiple factors, including advances in our understanding of biological systems, improvements in process control and monitoring capabilities, and the increasing demands for product quality and manufacturing efficiency.

The foundation of modern biomanufacturing rests upon the successful integration of multiple disciplines and approaches, from molecular biology and process engineering to quality systems and facility design. This integration has created sophisticated manufacturing frameworks that enable the consistent production of high-quality biological products while maintaining operational efficiency and regulatory compliance. The implementation of quality by design principles has fundamentally altered how biomanufacturing processes are developed and controlled, emphasizing the importance of systematic approaches to process understanding and control. This shift in methodology has led to more robust manufacturing processes that can consistently deliver high-quality products while maintaining operational efficiency.

The emergence of advanced technologies, particularly in the areas of process analytics and control, has revolutionized how biomanufacturing operations are monitored and managed. The integration of sophisticated monitoring systems with advanced control strategies has created new opportunities for process optimization and improvement, enabling manufacturers to achieve unprecedented levels of process understanding and control. The application of artificial intelligence and machine learning technologies represents the latest evolution in this technological progression, offering new tools for process optimization and control that were previously unavailable. These advanced analytical capabilities have enabled the identification of subtle process relationships and patterns that might not be apparent through traditional analysis methods, leading to improvements in process efficiency and product quality.

The life cycle approach to manufacturing has emerged as a critical framework for managing biological product manufacturing, emphasizing the importance of systematic approaches to process development, validation, and continuous improvement. This comprehensive view of manufacturing operations has created new opportunities for process optimization and improvement while ensuring consistent product quality throughout the product life cycle. The integration of knowledge management systems with manufacturing operations has enhanced our ability to capture and utilize critical process understanding, enabling more effective decision-making and process optimization throughout the product life cycle.

Looking toward the future, several key trends are likely to shape the continued evolution of biomanufacturing. The increasing adoption of continuous manufacturing technologies promises to revolutionize how biological products are manufactured, offering potential improvements in process efficiency and product quality. The integration of advanced analytics and artificial intelligence technologies will continue to enhance our ability to monitor and control manufacturing processes, leading to further improvements in process understanding and control. The development of more flexible manufacturing platforms will enable more efficient production of multiple products within the same facility, reducing manufacturing costs and improving facility utilization.

The regulatory landscape surrounding biomanufacturing will continue to evolve as new technologies and approaches emerge. The adoption of risk-based approaches to manufacturing oversight has created new opportunities for process innovation while maintaining the high standards of product quality and safety required for biological products. This evolution in regulatory approaches will likely continue as new manufacturing technologies and control strategies are developed and implemented. The increasing focus on sustainability in manufacturing operations will also influence how biomanufacturing processes are developed and implemented, leading to new approaches that minimize environmental impact while maintaining product quality and manufacturing efficiency.

The importance of personnel development and training will remain critical as biomanufacturing operations become increasingly complex. The successful implementation of advanced manufacturing technologies and approaches requires personnel with sophisticated technical skills and deep process understanding. Organizations must continue to invest in personnel development programs that ensure their workforce maintains the skills and knowledge needed to operate modern manufacturing facilities effectively. The integration of advanced training technologies, including virtual and augmented reality systems, may provide new opportunities for more effective personnel training and development.

The continued evolution of biomanufacturing will require ongoing collaboration between different stakeholders, from manufacturers and technology providers to regulatory authorities and academic institutions. This collaboration will be essential in addressing the challenges associated with implementing new manufacturing technologies and approaches while maintaining the high standards of product quality and safety required for biological products. The development of standard approaches to technology implementation and validation will facilitate the adoption of new manufacturing technologies while ensuring consistent product quality across different manufacturing facilities and organizations (Figure 1.11).

In conclusion, the field of biomanufacturing stands at an important junction, with new technologies and approaches creating opportunities for significant

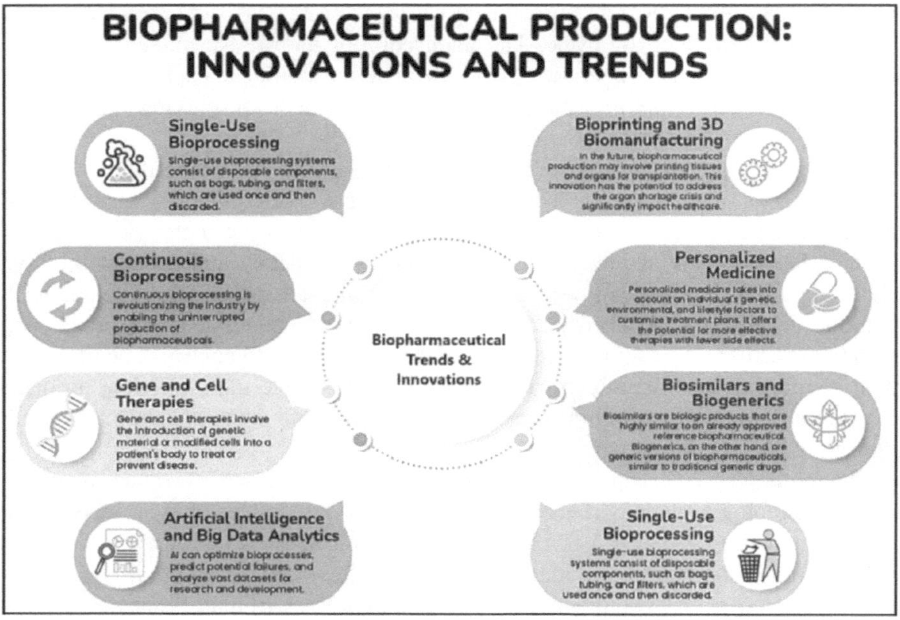

FIGURE 1.11 Future trends in biomanufacturing.

improvements in manufacturing efficiency and product quality. The successful integration of these advanced technologies with traditional manufacturing approaches will require careful consideration of multiple factors, from technical capabilities and regulatory requirements to personnel training and facility design. Organizations that can effectively navigate these challenges while maintaining focus on product quality and manufacturing efficiency will be well-positioned to succeed in the evolving landscape of biological product manufacturing. The future of biomanufacturing promises to be both challenging and exciting as new technologies and approaches continue to emerge, creating opportunities for further improvements in how biological products are manufactured and delivered to patients worldwide.

2 Raw Material and Compliance in Biomanufacturing

2.1 INTRODUCTION

The biomanufacturing industry has undergone significant transformation over the past decades, evolving from traditional pharmaceutical manufacturing processes to highly sophisticated, technology-driven production systems. This evolution has been particularly evident in the management of raw materials and compliance requirements, which have become increasingly complex and critical to the success of biomanufacturing operations. The integration of advanced technologies, coupled with stringent regulatory requirements, has created a new paradigm in how organizations approach raw material management and quality control in their biomanufacturing processes.

The significance of raw material management in biomanufacturing cannot be overstated, as it forms the foundation upon which product quality, process efficiency, and regulatory compliance are built. In modern biomanufacturing facilities, raw materials encompass a diverse range of components, including cell culture media, buffer solutions, chromatography resins, and various process additives. Each of these materials must be carefully selected, characterized, and monitored throughout the manufacturing process to ensure consistent product quality and regulatory compliance. The complexity of these materials, combined with their critical role in product quality, necessitates a comprehensive understanding of their properties, behaviors, and potential impacts on the manufacturing process.

The management of raw materials in biomanufacturing extends beyond simple inventory control and encompasses a complex web of activities including supplier qualification, material characterization, storage and handling procedures, and in-process controls. These activities must be conducted within a framework that ensures compliance with current Good Manufacturing Practices (cGMPs) and other relevant regulatory requirements. The increasing globalization of supply chains has added another layer of complexity to raw material management, requiring manufacturers to implement robust systems for supplier oversight and material traceability (Table 2.1).

Quality and compliance considerations in biomanufacturing have evolved significantly with the advancement of technology and increasing regulatory scrutiny. Modern biomanufacturing facilities must maintain robust quality management systems that ensure consistent compliance with regulatory requirements while

DOI: 10.1201/9781003624592-2

TABLE 2.1
Critical Aspects of Raw Material Management in Biomanufacturing

Aspect	Key Considerations	Impact on Manufacturing
Material selection	Chemical compatibility, biological activity, stability	Product quality, process consistency
Supplier management	Qualification, audit programs, quality agreements	Supply chain reliability, compliance
Storage conditions	Temperature, humidity, containment	Material stability, contamination prevention
Quality control	Testing protocols, specifications, trending	Process control, product consistency

supporting efficient operations. This includes implementing comprehensive documentation systems, conducting regular audits, and maintaining ongoing training programs for personnel involved in raw material handling and management.

The introduction of lean manufacturing principles into biomanufacturing operations has created new opportunities for optimizing raw material management while maintaining strict quality and compliance standards. These principles emphasize the elimination of waste, continuous improvement of processes, and the creation of value for end users. However, the application of lean principles in biomanufacturing must be carefully balanced against the need to maintain robust quality controls and regulatory compliance.

The emergence of artificial intelligence (AI) and advanced analytics has introduced new possibilities for enhancing raw material management in biomanufacturing. These technologies enable more sophisticated approaches to quality prediction, supply chain optimization, and process control. The integration of AI-driven systems with traditional quality control methods has created new paradigms for ensuring consistent product quality while improving operational efficiency.

The growing importance of sustainability in manufacturing has also influenced how organizations approach raw material management in biomanufacturing. This includes considerations such as material sourcing, waste reduction, and environmental impact. Manufacturers must now balance these sustainability goals with traditional objectives of quality, efficiency, and compliance, creating new challenges and opportunities in raw material management.

This chapter explores these various aspects of raw material management and compliance in biomanufacturing, providing a comprehensive examination of current practices, emerging technologies, and future trends. The discussion encompasses both theoretical frameworks and practical applications, offering insights into how organizations can effectively manage raw materials while maintaining compliance with regulatory requirements and achieving operational excellence.

2.2 RAW MATERIAL CONSIDERATIONS IN BIOMANUFACTURING

2.2.1 Fundamental Principles of Raw Material Selection

The selection of raw materials for biomanufacturing processes represents a complex decision-making process that requires careful consideration of multiple factors including material quality, consistency, availability, and cost-effectiveness. In modern biomanufacturing facilities, raw material selection begins with a comprehensive understanding of the intended manufacturing process and the specific requirements of the final product. This includes detailed analysis of the material's chemical and biological properties, their potential interactions with other process components, and their impact on the final product quality. The selection process must also consider the material's stability under various processing conditions, its shelf life, and any special handling or storage requirements that might affect its usability in the manufacturing process.

Material characterization plays a pivotal role in the selection process, requiring extensive analytical testing to ensure that materials meet predetermined specifications. This characterization process extends beyond simple chemical analysis to include detailed examination of physical properties, biological activity, and potential impurities. The understanding gained through this characterization process forms the foundation for establishing material specifications and developing appropriate control strategies. These specifications must be sufficiently detailed to ensure consistent material quality while being realistic enough to be consistently achievable by suppliers.

The impact of raw material variability on process performance and product quality cannot be overstated. Even minor variations in raw material characteristics can lead to significant changes in process performance and product quality. Therefore, manufacturers must establish robust systems for monitoring and controlling raw material variability. This includes implementing appropriate sampling plans, conducting trend analysis of material properties, and maintaining detailed records of material performance in actual manufacturing processes (Figure 2.1).

2.2.2 Material Characterization and Specification Development

The development of raw material specifications represents a critical element of the overall quality control strategy in biomanufacturing. These specifications must be based on a thorough understanding of the material's critical quality attributes and their potential impact on process performance and product quality. The specification development process involves extensive analytical testing, process validation studies, and risk assessments to identify appropriate acceptance criteria for each material parameter.

Material characterization studies play a fundamental role in understanding how various raw material attributes influence process performance and product quality. These studies typically involve detailed analytical testing using multiple complementary techniques to evaluate chemical composition, physical properties, and biological

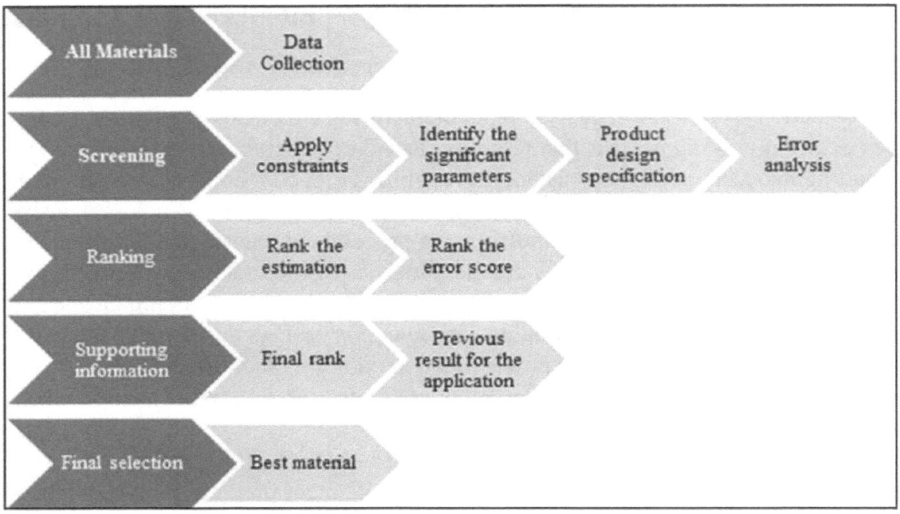

FIGURE 2.1 Raw material selection and characterization framework.

activity. The results of these studies inform the development of appropriate specifications and testing requirements for incoming materials as well as the establishment of appropriate storage conditions and handling procedures (Figure 2.2).

The establishment of meaningful specifications requires careful consideration of analytical method capabilities, process requirements, and regulatory expectations. Organizations must balance the need for tight specifications to ensure consistent product quality against practical considerations such as analytical variability, supplier capabilities, and cost implications. This often involves extensive collaboration between quality control, manufacturing, and supplier organizations to develop specifications that are both meaningful and achievable.

2.2.3 Supply Chain Considerations for Raw Materials

The globalization of biomanufacturing supply chains has introduced new complexities into raw material management. Manufacturers must now consider factors such as supplier location, transportation requirements, and international regulations when selecting raw materials. The establishment of reliable supply chains requires careful evaluation of potential suppliers, including assessment of their quality systems, manufacturing capabilities, and compliance with relevant regulatory requirements. This evaluation process often involves detailed audits of supplier facilities, review of quality documentation, and ongoing monitoring of supplier performance.

Supplier qualification represents a critical aspect of raw material management in biomanufacturing. The qualification process typically involves multiple stages, beginning with initial supplier assessment and progressing through various levels of evaluation before a supplier is approved for use. This includes review of the supplier's

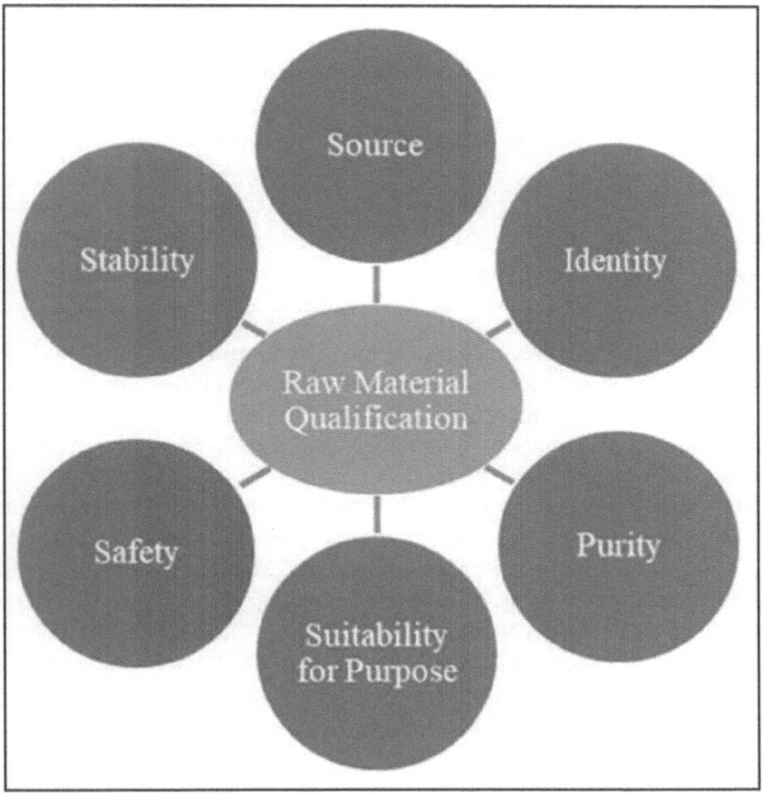

FIGURE 2.2 Raw material quality impact framework.

quality systems, manufacturing processes, and compliance history. Once qualified, suppliers must be continuously monitored to ensure they maintain acceptable performance levels and comply with established quality agreements.

Risk management in raw material supply chains has become increasingly important as manufacturing operations become more global. Manufacturers must develop comprehensive risk assessment and mitigation strategies to address potential supply chain disruptions. This includes establishing backup suppliers for critical materials, maintaining appropriate inventory levels, and developing contingency plans for addressing supply chain interruptions. The implementation of effective risk management strategies requires close collaboration between quality, manufacturing, and supply chain functions within the organization.

2.2.4 Storage and Handling Requirements

The proper storage and handling of raw materials represents a critical aspect of biomanufacturing operations. Materials must be stored under controlled conditions that

TABLE 2.2
Critical Storage Parameters for Common Raw Material Categories

Material Category	Temperature Requirements	Humidity Controls	Special Considerations
Cell culture media	2–8°C	N/A	Light protection required
Buffer components	15–25°C	<60% RH	Moisture sensitive
Chromatography resins	2–8°C	N/A	No freezing allowed
Process additives	15–25°C	<40% RH	Chemical compatibility

maintain their quality and prevent degradation or contamination. This requires careful consideration of environmental factors such as temperature, humidity, and light exposure. Storage facilities must be designed and maintained to ensure appropriate environmental control while providing adequate space for proper material segregation and handling.

Material handling procedures must be carefully designed to prevent contamination and ensure material integrity throughout the manufacturing process. This includes establishing appropriate procedures for material receipt, sampling, dispensing, and transfer between different areas of the facility. Personnel involved in material handling must be properly trained in these procedures and must demonstrate competency in their execution. The implementation of effective material handling systems requires careful consideration of facility layout, equipment design, and personnel movement patterns (Table 2.2).

2.2.5 Quality Control and Testing Requirements

Quality control of raw materials extends beyond initial receipt testing to include ongoing monitoring throughout the material's life cycle. This includes establishment of appropriate sampling plans, development of test methods, and implementation of trend analysis programs. The testing strategy must be designed to detect both obvious defects and subtle changes in material properties that might affect process performance or product quality. This requires careful selection of test methods that are both appropriately sensitive and sufficiently robust for routine use.

The implementation of effective testing programs requires not only appropriate analytical methods but also robust systems for sample management, data analysis, and trending. Organizations must establish appropriate procedures for sample collection, handling, and storage to ensure that testing results accurately reflect material quality. The development of meaningful specification limits requires careful consideration of method capabilities, process requirements, and historical data (Table 2.3).

2.2.6 Documentation and Traceability Systems

The implementation of effective documentation and traceability systems is essential for proper raw material management in biomanufacturing. These systems must

TABLE 2.3
Raw Material Testing Requirements in Biomanufacturing

Test Category	Purpose	Frequency	Critical Parameters
Identity	Material verification	Every container	Chemical/physical markers
Purity	Contamination detection	Based on risk	Impurity profiles
Potency	Activity confirmation	Lot specific	Biological activity
Stability	Degradation monitoring	Time based	Critical attributes

provide complete documentation of material receipt, testing, release, and use in manufacturing operations. This includes maintaining detailed records of material specifications, test results, and any investigations or deviations that occur during material handling and use. The documentation system must support both internal quality requirements and regulatory compliance needs while providing sufficient detail to enable effective material tracking and traceability.

2.2.7 Supplier Quality Management

The management of supplier quality represents a critical element of raw material control in biomanufacturing. This includes not only initial supplier qualification activities but also ongoing monitoring and oversight of supplier performance. Organizations must develop comprehensive programs for evaluating and qualifying suppliers, including technical assessments, quality system audits, and review of regulatory compliance history.

2.2.8 Material Risk Assessment and Control Strategies

Risk assessment and control strategies for raw materials must consider both the potential impact of material variability on process performance and product quality and the likelihood of encountering such variability. This requires development of comprehensive risk assessment tools that consider multiple factors including material criticality, supply chain reliability, and potential impact on product quality. The results of these risk assessments must be used to develop appropriate control strategies that ensure consistent material quality while maintaining operational efficiency.

The development of appropriate risk mitigation strategies requires careful consideration of both technical and operational factors. This includes implementation of appropriate controls at various stages of the supply chain, from initial material procurement through final use in manufacturing. Organizations must also establish appropriate monitoring systems to ensure the effectiveness of risk mitigation measures and identify potential issues before they impact production operations.

2.3 COMPLIANCE AND QUALITY IN BIOMANUFACTURING

2.3.1 Regulatory Framework and Quality Systems Integration

The foundation of compliance in biomanufacturing rests upon a complex interplay of regulatory requirements, industry standards, and organizational quality systems. Modern biomanufacturing facilities must navigate an intricate web of regulations from multiple authorities, including the FDA, EMA, and other regional regulatory bodies. These requirements not only govern the final product quality but also extend to every aspect of the manufacturing process, from raw material handling to equipment validation and personnel training. The implementation of these regulatory requirements demands a systematic approach that integrates quality considerations into every aspect of the manufacturing operation, creating a comprehensive framework that ensures consistent compliance while supporting efficient production.

Quality systems in biomanufacturing have evolved significantly from their origins in traditional pharmaceutical manufacturing. Contemporary quality systems must address the unique challenges presented by biological products, including process variability, product stability, and the complexity of biological systems. This evolution has led to the development of sophisticated quality management systems that incorporate risk-based approaches, continuous monitoring, and advanced analytical methods. Organizations must establish robust documentation systems that capture not only the results of quality control testing but also process parameters, environmental conditions, and personnel activities that could impact product quality.

The integration of quality systems with operational processes requires careful consideration of organizational structure, communication pathways, and decision-making processes. Successful implementation depends on creating clear lines of responsibility, establishing effective communication channels, and fostering a culture of quality throughout the organization. This integration must extend beyond traditional quality control functions to encompass all aspects of the operation, including maintenance, facilities management, and supply chain operations (Figure 2.3).

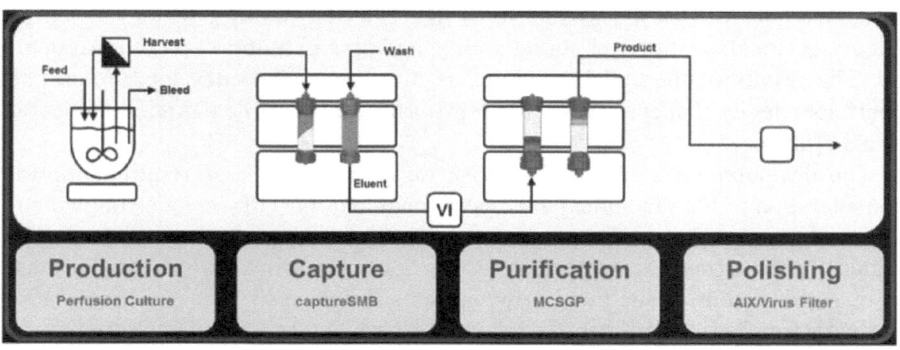

FIGURE 2.3 Integrated quality system architecture in biomanufacturing.

2.3.2 Risk Management and Quality Control Strategies

Risk management in biomanufacturing represents a critical component of the quality system, requiring a comprehensive approach that identifies, evaluates, and mitigates potential risks to product quality and regulatory compliance. Organizations must develop sophisticated risk assessment methodologies that consider not only direct process risks but also indirect factors such as supply chain disruptions, environmental conditions, and personnel capabilities. The implementation of effective risk management strategies requires continuous monitoring and assessment of both internal and external factors that could impact product quality or regulatory compliance.

Quality control strategies in biomanufacturing must address the inherent variability of biological processes while maintaining consistent product quality. This requires the development of robust testing protocols, specification limits, and trending analyses that can detect potential quality issues before they impact product quality. Modern quality control approaches incorporate advanced analytical methods, including real-time monitoring systems and rapid microbial detection techniques, to provide timely information about process performance and product quality (Table 2.4).

2.3.3 Documentation and Data Management Systems

Documentation systems in biomanufacturing must provide comprehensive evidence of compliance while supporting efficient operations. Modern documentation approaches utilize electronic systems that facilitate real-time data capture, automated workflows, and integrated quality management functions. These systems must maintain data integrity while providing accessibility to authorized personnel and supporting regulatory review processes. The implementation of electronic documentation systems requires careful consideration of system validation, data security, and backup procedures to ensure continued compliance with regulatory requirements.

Effective data management in biomanufacturing extends beyond simple record-keeping to encompass sophisticated analysis and trending capabilities. Organizations

TABLE 2.4
Quality Control Strategy Components in Biomanufacturing

Component	Key Elements	Implementation Considerations
In-process testing	Critical parameter monitoring, sample analysis	Test method validation, data trending
Environmental monitoring	Facility conditions, contamination control	Alert/action limits, response procedures
Final product testing	Release specifications, stability studies	Method qualification, reference standards
Documentation	Batch records, test results, investigations	Data integrity, record retention

must develop systems that can process large volumes of manufacturing data, identify trends and patterns, and support quality decision-making processes. This includes implementing data governance procedures that ensure data quality, consistency, and reliability throughout the manufacturing operation.

2.3.4 Validation and Qualification Programs

Validation programs in biomanufacturing must address all aspects of the manufacturing process, including equipment, facilities, utilities, and computerized systems. The development of effective validation strategies requires a thorough understanding of process requirements, regulatory expectations, and industry best practices. Organizations must establish comprehensive validation master plans that outline the approach to validation, including requirements for installation qualification, operational qualification, and performance qualification activities.

The qualification of manufacturing equipment and systems represents a critical component of the validation program. This includes developing detailed protocols for equipment qualification, establishing acceptance criteria for critical parameters, and documenting the results of qualification activities. The qualification program must also address ongoing requirements for preventive maintenance, calibration, and periodic requalification to ensure continued compliance with regulatory requirements.

2.3.5 Personnel Training and Competency Assessment

Personnel training programs in biomanufacturing must ensure that all employees possess the knowledge and skills required to perform their assigned duties effectively while maintaining compliance with regulatory requirements. This includes developing comprehensive training curricula that address both technical skills and regulatory requirements, implementing effective assessment methods to verify competency, and maintaining detailed records of training activities.

The development of effective training programs requires consideration of various learning styles, regulatory requirements, and operational needs. Organizations must establish systems for identifying training needs, developing appropriate training materials, and evaluating the effectiveness of training programs. This includes implementing competency assessment programs that verify employees' ability to perform their assigned duties consistently and in compliance with established procedures.

2.3.6 Supplier Quality Management

Supplier quality management in biomanufacturing requires a comprehensive approach that ensures the consistent quality of materials and services while maintaining regulatory compliance. Organizations must develop robust supplier qualification programs that include initial assessment, ongoing monitoring, and periodic requalification of critical suppliers. This includes establishing quality agreements that clearly define responsibilities, quality requirements, and communication protocols between the manufacturer and supplier.

The implementation of effective supplier management programs requires consideration of various factors, including supplier capabilities, geographic location, and regulatory requirements. Organizations must establish systems for monitoring supplier performance, conducting quality audits, and managing supplier-related quality issues. This includes developing procedures for investigating and resolving supplier-related quality problems while maintaining appropriate documentation of all supplier management activities.

2.4 LEAN BIOMANUFACTURING

2.4.1 Fundamentals of Lean Implementation in Biomanufacturing

The implementation of lean principles in biomanufacturing represents a paradigm shift from traditional manufacturing approaches, requiring careful adaptation to meet the unique challenges of biological production systems. Unlike conventional manufacturing, where lean principles were originally developed, biomanufacturing involves complex biological processes that demand specialized considerations for successful lean implementation. The fundamental aspects of lean manufacturing, including value stream mapping, waste reduction, and continuous flow, must be carefully tailored to accommodate the inherent variability and complexity of biological processes while maintaining product quality and regulatory compliance.

The adaptation of lean principles to biomanufacturing environments requires a comprehensive understanding of both manufacturing efficiency and biological process requirements. Organizations must carefully balance the desire for process optimization with the need to maintain strict control over critical process parameters and product quality attributes. This balance becomes particularly crucial when considering the implementation of just-in-time inventory systems for sensitive biological materials, where storage conditions and material stability play vital roles in maintaining product quality.

The successful implementation of lean principles in biomanufacturing necessitates a cultural transformation within the organization. This transformation extends beyond simple process improvements to encompass changes in organizational structure, decision-making processes, and employee engagement. The development of a lean culture in biomanufacturing requires extensive training programs, clear communication channels, and strong leadership commitment to support the transition from traditional manufacturing approaches to lean methodologies (Figure 2.4).

2.4.2 Value Stream Optimization in Biological Processes

Value stream optimization in biomanufacturing presents unique challenges due to the complex nature of biological processes and the stringent regulatory requirements governing pharmaceutical production. The identification and elimination of non-value-adding activities must be conducted with careful consideration of

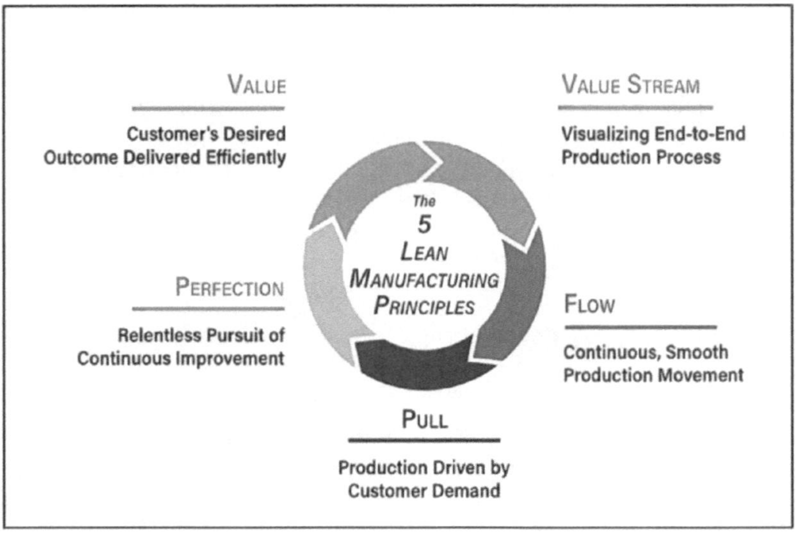

FIGURE 2.4 Integrated lean biomanufacturing framework.

their potential impact on product quality and process stability. This requires a detailed understanding of process interdependencies and the development of sophisticated approaches to value stream mapping that account for biological process variability.

The optimization of value streams in biomanufacturing extends beyond traditional manufacturing considerations to encompass unique aspects such as cell culture processes, purification operations, and analytical testing requirements. Each of these elements must be carefully evaluated to identify opportunities for efficiency improvements while maintaining the integrity of the biological processes and ensuring compliance with regulatory requirements. The development of optimized value streams often requires the integration of advanced monitoring systems and real-time process control strategies to maintain process stability and product quality (Table 2.5).

TABLE 2.5
Value Stream Optimization Parameters in Biomanufacturing

Parameter Category	Optimization Approach	Impact on Process	Quality Considerations
Process flow	Continuous processing	Reduced cycle time	Product stability
Material movement	Just-in-time delivery	Inventory reduction	Material integrity
Quality control	Real-time monitoring	Faster release	Compliance assurance
Equipment usage	Preventive maintenance	Higher efficiency	Process reliability

2.4.3 Waste Reduction Strategies in Biological Production

The concept of waste reduction in biomanufacturing encompasses multiple dimensions beyond traditional manufacturing waste. These include the management of biological materials, the optimization of complex buffer preparation processes, and the reduction of analytical testing waste. The implementation of waste reduction strategies must carefully consider the impact on product quality and process reliability while maintaining compliance with regulatory requirements governing pharmaceutical manufacturing.

The development of effective waste reduction strategies requires a comprehensive understanding of biological process requirements and the potential impact of process changes on product quality. Organizations must carefully evaluate opportunities for waste reduction against the need to maintain robust process control and product quality assurance. This evaluation often involves sophisticated risk assessment approaches that consider both operational efficiency and product quality implications.

2.4.4 Continuous Improvement in Biological Processing

Continuous improvement in biomanufacturing environments requires specialized approaches that account for the unique characteristics of biological processes. The implementation of continuous improvement initiatives must consider factors such as process variability, material stability, and regulatory compliance requirements. This necessitates the development of sophisticated monitoring systems and data analysis approaches to identify improvement opportunities while maintaining process control and product quality.

The establishment of effective continuous improvement programs in biomanufacturing requires the integration of multiple organizational functions, including process development, manufacturing operations, quality control, and regulatory compliance. These functions must work collaboratively to identify and implement improvements while ensuring that changes do not adversely impact product quality or regulatory compliance status. The development of cross-functional improvement teams and robust change management processes becomes crucial for successful continuous improvement implementation.

2.4.5 Quality Integration in Lean Biomanufacturing

The integration of quality systems within lean biomanufacturing operations represents a critical aspect of successful implementation. Organizations must develop approaches that maintain robust quality control while supporting lean operational objectives. This integration requires careful consideration of regulatory requirements, product quality attributes, and operational efficiency goals to ensure that lean implementation does not compromise product quality or compliance status.

The development of integrated quality systems in lean biomanufacturing environments requires sophisticated approaches to quality risk management and process control. Organizations must establish systems that enable rapid detection and response to quality issues while maintaining efficient operations. This often involves the implementation of advanced monitoring systems and real-time quality control approaches that support both quality assurance and operational efficiency objectives.

2.4.6 Performance Measurement and Monitoring

The establishment of effective performance measurement systems in lean biomanufacturing environments requires careful consideration of both operational and quality metrics. Organizations must develop comprehensive measurement approaches that capture key performance indicators while maintaining focus on critical quality attributes and compliance requirements. This necessitates the integration of multiple data sources and the development of sophisticated analysis approaches to support decision-making processes.

The implementation of performance monitoring systems must consider the unique characteristics of biological processes and their impact on operational performance. Organizations must establish monitoring approaches that account for process variability while providing meaningful insights into operational efficiency and quality performance. The development of balanced scorecard approaches that integrate operational, quality, and compliance metrics becomes crucial for effective performance management in lean biomanufacturing environments.

2.5 ARTIFICIAL INTELLIGENCE IN RAW MATERIAL MANAGEMENT

2.5.1 Fundamentals of AI Implementation in Biomanufacturing

The integration of artificial intelligence into raw material management represents a paradigm shift in how biomanufacturing facilities approach quality control and process optimization. Traditional methods of raw material management relied heavily on human expertise and manual intervention, often resulting in inefficiencies and potential inconsistencies in decision-making processes. The implementation of AI systems has revolutionized this approach by introducing automated, data-driven decision-making capabilities that can process vast amounts of information in real time, identifying patterns and trends that might be impossible for human operators to detect.

The foundation of AI implementation in raw material management lies in the sophisticated integration of multiple data sources, including supplier information, quality control metrics, environmental monitoring data, and process parameters. These diverse data streams are consolidated into comprehensive databases that serve as the foundation for AI-driven analytics. The complexity of biomanufacturing processes, combined with the critical nature of raw material quality, necessitates an approach that can handle multiple variables simultaneously while maintaining the highest standards of accuracy and reliability.

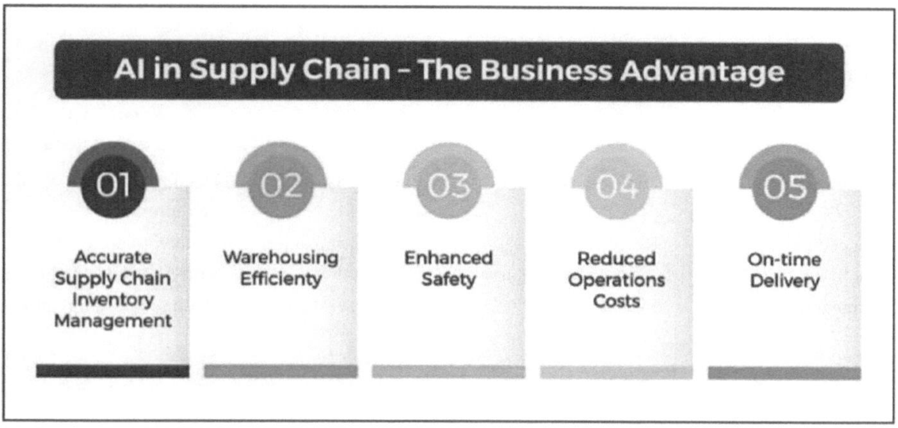

FIGURE 2.5 AI integration framework for raw material management.

2.5.2 Machine Learning Algorithms in Quality Assessment

The application of machine learning algorithms in raw material quality assessment has transformed the traditional approach to quality control in biomanufacturing. These algorithms utilize sophisticated pattern recognition capabilities to identify subtle variations in raw material characteristics that might impact product quality. The development of these systems requires extensive training using historical data, where algorithms learn to recognize patterns associated with both acceptable and suboptimal material quality (Figure 2.5).

Machine learning models employed in raw material management typically incorporate multiple layers of analysis, examining not only the direct quality parameters but also the relationships between different material characteristics and their collective impact on process outcomes. This comprehensive approach enables more accurate predictions of material performance and potential quality issues before they manifest in the production process.

2.5.3 Real-Time Monitoring and Adaptive Control Systems

The implementation of AI-driven real-time monitoring systems has enabled biomanufacturing facilities to maintain continuous oversight of raw material quality throughout the production process. These systems utilize advanced sensors and data collection mechanisms to gather information about material properties and process conditions continuously. The integration of real-time monitoring with adaptive control systems allows for immediate responses to any deviations from optimal conditions, ensuring consistent product quality while minimizing waste and process interruptions.

The development of adaptive control systems represents a significant advancement in raw material management, as these systems can automatically adjust process

TABLE 2.6
AI Implementation Levels in Raw Material Management

Implementation Level	Key Features	Benefits	Regulatory Considerations
Basic analytics	Pattern recognition	Improved quality control	Documentation requirements
Advanced monitoring	Real-time analysis	Process optimization	Validation protocols
Predictive control	Automated response	Risk reduction	Change management
Full integration	System-wide optimization	Maximum efficiency	Continuous compliance

parameters based on real-time analysis of material characteristics and process conditions. This capability ensures optimal utilization of raw materials while maintaining strict quality standards and regulatory compliance. The implementation of such systems requires careful consideration of both technical capabilities and regulatory requirements, ensuring that automated decisions align with established quality control protocols (Table 2.6).

2.5.4 Data Integration and Knowledge Management

The success of AI implementation in raw material management heavily depends on effective data integration and knowledge management strategies. Modern biomanufacturing facilities must develop robust systems for collecting, storing, and analyzing data from multiple sources, ensuring that information is both accessible and secure. The development of comprehensive data management systems requires careful consideration of data quality, system integration, and regulatory compliance requirements.

The implementation of knowledge management systems alongside AI capabilities ensures that institutional knowledge and expertise are effectively captured and utilized in decision-making processes. These systems enable the preservation and application of critical knowledge, ensuring consistency in raw material management practices across different production runs and facilities.

2.5.5 Risk Assessment and Mitigation Strategies

The integration of AI in raw material management introduces new considerations for risk assessment and mitigation. While AI systems offer significant benefits in terms of process optimization and quality control, they also require careful evaluation of potential risks associated with automated decision-making. The development of comprehensive risk assessment frameworks specific to AI implementation ensures that potential issues are identified and addressed proactively.

Risk mitigation strategies in AI-driven raw material management must address both technical and regulatory considerations. This includes developing robust validation protocols for AI systems, establishing clear procedures for system monitoring and maintenance, and ensuring appropriate oversight of automated decision-making processes. The implementation of these strategies requires close collaboration between technical experts, quality assurance personnel, and regulatory compliance teams.

2.5.6 Future Trends and Technological Advancements

The evolution of AI technologies continues to create new opportunities for enhancing raw material management in biomanufacturing. Emerging technologies such as quantum computing and advanced neural networks offer the potential for even more sophisticated analysis and control capabilities. The development of these technologies will likely lead to further improvements in process efficiency and product quality while potentially reducing the complexity of regulatory compliance.

The integration of AI with other emerging technologies, such as blockchain and Internet of Things (IoT) devices, presents opportunities for enhanced traceability and security in raw material management. These technological combinations could enable more comprehensive oversight of supply chains, improved quality assurance, and more efficient regulatory compliance processes. The continued evolution of these technologies will likely shape the future of raw material management in biomanufacturing, creating new possibilities for process optimization and quality control.

2.6 PREDICTIVE ANALYTICS FOR RAW MATERIAL QUALITY AND SUPPLY CHAIN OPTIMIZATION

2.6.1 Foundations of Predictive Analytics in Biomanufacturing

The implementation of predictive analytics in biomanufacturing represents a paradigm shift in how organizations approach raw material quality management and supply chain optimization. This sophisticated approach combines statistical methods, machine learning algorithms, and domain expertise to forecast potential quality issues and supply chain disruptions before they materialize. The fundamental premise of predictive analytics in this context lies in its ability to process vast amounts of historical and real-time data to generate actionable insights that can guide decision-making processes throughout the biomanufacturing value chain.

In modern biomanufacturing facilities, the application of predictive analytics extends far beyond simple trend analysis. It encompasses a comprehensive framework that integrates data from multiple sources, including supplier quality metrics, environmental monitoring systems, process parameters, and finished product specifications. This integration creates a holistic view of the manufacturing process, enabling organizations to identify subtle patterns and correlations that might otherwise go unnoticed. The sophistication of these systems has evolved to the point

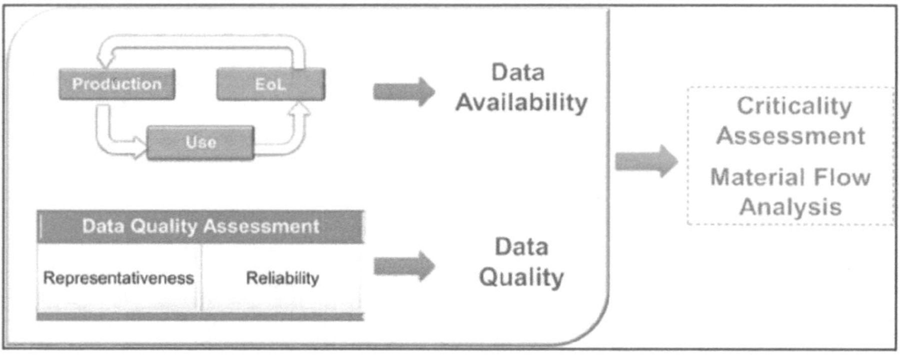

FIGURE 2.6 Integrated predictive analytics framework for raw material quality management.

where they can now detect potential quality deviations weeks or even months before they would become apparent through traditional quality control methods.

The evolution of predictive analytics in biomanufacturing has been driven by several key factors, including the increasing availability of high-quality data, advancements in computing power, and the development of more sophisticated algorithmic approaches. These developments have enabled organizations to move from reactive quality control measures to proactive quality assurance strategies that can anticipate and prevent quality issues before they impact production. This shift has profound implications for both operational efficiency and product quality, as it allows manufacturers to optimize their processes in real time while maintaining strict compliance with regulatory requirements (Figure 2.6).

2.6.2 Advanced Data Integration and Analysis Methods

The successful implementation of predictive analytics in raw material management requires sophisticated approaches to data integration and analysis. Modern biomanufacturing facilities generate enormous amounts of data from various sources, including in-process controls, environmental monitoring systems, and supplier quality metrics. The challenge lies not only in collecting this data but also in integrating it in a way that provides meaningful insights for decision-making processes. Advanced data integration methods must account for different data formats, sampling frequencies, and quality levels while ensuring that the integrated dataset maintains its statistical validity and relevance to the manufacturing process.

Data analysis in this context goes beyond traditional statistical methods to incorporate advanced techniques such as multivariate analysis, pattern recognition, and machine learning algorithms. These methods allow organizations to identify complex relationships between raw material attributes and process outcomes that might not be apparent through conventional analysis. The sophistication of these analytical approaches has increased significantly in recent years, with the development of new algorithms specifically designed to handle the unique challenges of biomanufacturing data.

Effective data integration also requires careful consideration of data quality and validation procedures. Organizations must implement robust systems for data verification and cleaning to ensure that the insights generated through predictive analytics are based on accurate and reliable information. This includes developing standardized procedures for handling missing data, identifying and correcting outliers, and validating the accuracy of sensor readings and other automated data collection systems.

2.6.3 Supply Chain Network Optimization

The application of predictive analytics to supply chain optimization represents another critical aspect of raw material management in biomanufacturing. Modern supply chains are complex networks that must balance multiple competing objectives, including cost efficiency, reliability, and sustainability. Predictive analytics provides organizations with the tools needed to optimize these networks while maintaining the high levels of quality and reliability required in biomanufacturing.

Supply chain optimization through predictive analytics involves analyzing historical supply chain performance data, market trends, and external factors to identify potential risks and opportunities. This analysis enables organizations to make more informed decisions about supplier selection, inventory management, and logistics planning. The sophistication of these systems has evolved to the point where they can now predict potential supply chain disruptions with remarkable accuracy, allowing organizations to take preemptive action to mitigate risks.

One of the most significant advantages of predictive analytics in supply chain optimization is its ability to account for multiple variables simultaneously. Traditional approaches to supply chain management often relied on simplified models that could only consider a limited number of factors. Modern predictive analytics systems can process dozens or even hundreds of variables simultaneously, providing a much more comprehensive view of supply chain performance and potential optimization opportunities.

2.6.4 Quality Risk Assessment and Mitigation

The application of predictive analytics to quality risk assessment represents a significant advancement in how organizations approach quality management in biomanufacturing. Traditional quality control methods often relied on reactive approaches that could only identify quality issues after they had occurred. Predictive analytics enables a more proactive approach to quality management by identifying potential quality risks before they materialize into actual problems (Table 2.7).

2.6.5 Implementation Strategies and Challenges

The implementation of predictive analytics systems in biomanufacturing presents several unique challenges that organizations must address. These challenges range from technical considerations, such as data integration and system validation,

TABLE 2.7
Risk Assessment Parameters in Predictive Analytics

Risk Category	Predictive Indicators	Mitigation Strategies	Implementation Timeline
Material quality	Trend analysis, stability predictions	Enhanced testing protocols	Short term
Supply chain	Supplier performance metrics	Diversification strategies	Medium term
Process control	Real-time monitoring data	Automated adjustments	Immediate

to organizational factors like change management and training requirements. Successful implementation requires a carefully planned approach that addresses these challenges while maintaining compliance with regulatory requirements and quality standards.

One of the most significant challenges in implementing predictive analytics systems is ensuring that they can be validated according to regulatory requirements. This includes developing appropriate validation protocols, establishing system boundaries, and implementing controls to ensure data integrity and security. Organizations must also consider how these systems will integrate with existing quality management systems and standard operating procedures.

2.6.6 Future Directions and Emerging Technologies

The future of predictive analytics in biomanufacturing continues to evolve with the development of new technologies and analytical approaches. Emerging technologies such as quantum computing and advanced artificial intelligence systems promise to further enhance the capabilities of predictive analytics systems. These developments could enable even more sophisticated approaches to quality prediction and supply chain optimization, potentially revolutionizing how organizations approach raw material management in biomanufacturing.

The integration of predictive analytics with other emerging technologies, such as blockchain and Internet of Things (IoT) devices, presents additional opportunities for enhancing raw material management and supply chain optimization. These technologies could provide new ways to track and verify raw material quality throughout the supply chain while enabling more sophisticated approaches to real-time process optimization and quality control.

2.7 ADVANCED TECHNOLOGIES AND FUTURE DIRECTIONS IN BIOMANUFACTURING

2.7.1 Integration of Smart Manufacturing Systems

The evolution of biomanufacturing systems has reached a critical juncture with the integration of smart manufacturing technologies that fundamentally transform how raw materials are managed and processed. Smart manufacturing systems in biomanufacturing environments represent a convergence of physical production processes with digital technologies, creating an interconnected ecosystem that enables unprecedented levels of control, monitoring, and optimization. These systems leverage advanced sensors, real-time data analytics, and automated control mechanisms to maintain optimal processing conditions while ensuring consistent product quality. The implementation of smart manufacturing principles in biomanufacturing facilities has led to significant improvements in process efficiency, reduced variability in product quality, and enhanced compliance with regulatory requirements.

The foundation of smart manufacturing in biomanufacturing lies in the sophisticated network of sensors and monitoring systems that continuously track critical process parameters. These systems generate massive amounts of data that provide insights into process performance, material behavior, and potential quality issues. The integration of these monitoring systems with advanced control algorithms enables real-time adjustments to process parameters, ensuring optimal conditions are maintained throughout the manufacturing process. This level of control and monitoring represents a significant advancement over traditional manufacturing approaches, where process adjustments were often reactive rather than proactive.

Smart manufacturing systems also facilitate the implementation of predictive maintenance strategies, reducing unexpected equipment failures and associated production interruptions. By analyzing patterns in equipment performance data, these systems can identify potential issues before they lead to failures, allowing maintenance activities to be scheduled during planned downtime periods. This predictive approach to maintenance has proven particularly valuable in biomanufacturing, where equipment reliability is critical to maintaining product quality and meeting production schedules.

The integration of smart manufacturing systems has also revolutionized how organizations manage their raw material inventory and supply chains. Advanced inventory management systems use real-time consumption data and predictive analytics to optimize stock levels and automatically trigger reordering processes when supplies reach predetermined thresholds. These systems consider factors such as lead times, usage patterns, and seasonal variations to ensure adequate material availability while minimizing excess inventory (Figure 2.7).

2.7.2 Advanced Process Analytical Technology Implementation

Process analytical technology (PAT) has emerged as a cornerstone of modern biomanufacturing operations, providing real-time insights into process performance and product quality. The implementation of advanced PAT systems represents a significant shift from traditional quality control approaches, where product quality was primarily assessed through end-point testing. Modern PAT systems enable continuous monitoring of critical quality attributes throughout the manufacturing process, allowing for immediate detection and correction of process deviations.

The evolution of PAT in biomanufacturing has been driven by advances in analytical instrumentation and data processing capabilities. Modern PAT systems

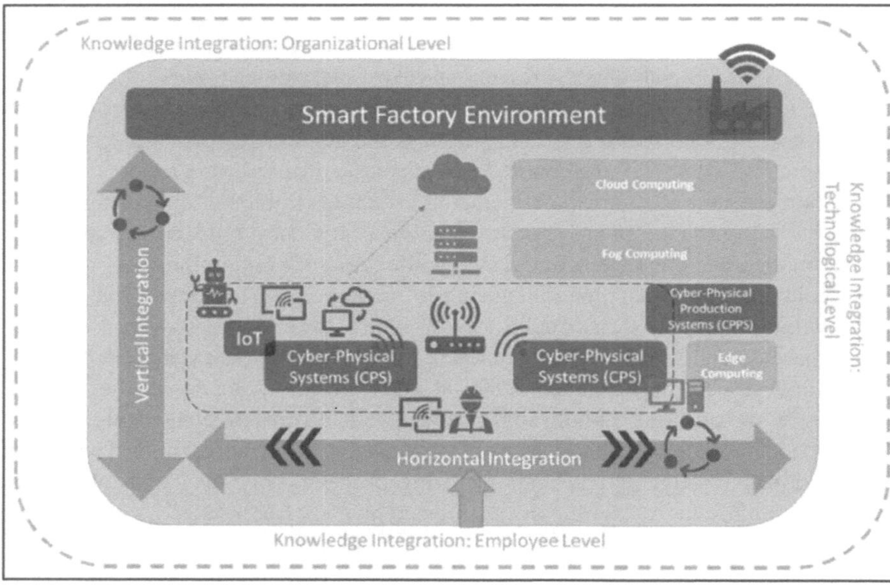

FIGURE 2.7 Smart manufacturing integration framework in biomanufacturing.

incorporate sophisticated spectroscopic techniques, including near-infrared and Raman spectroscopy, which provide detailed information about material properties and process conditions without requiring sample extraction or preparation. These noninvasive monitoring techniques have proven particularly valuable in maintaining the sterility of bioprocesses while providing continuous quality assurance.

The integration of PAT with process control systems has enabled the development of real-time release testing protocols, where product quality is assured through comprehensive in-process monitoring rather than relying solely on end-product testing. This approach not only reduces the time required for product release but also provides greater assurance of product quality through continuous monitoring throughout the manufacturing process. The implementation of real-time release testing represents a significant advancement in quality assurance practices and has been increasingly accepted by regulatory authorities as a valid approach to product quality verification (Table 2.8).

2.7.3 Artificial Intelligence and Machine Learning Applications

The integration of artificial intelligence and machine learning technologies in biomanufacturing has created new opportunities for process optimization and quality control. These technologies enable the analysis of complex data patterns that would be impossible to identify through traditional statistical methods. The application of AI in biomanufacturing extends across multiple areas, including process optimization, quality prediction, and maintenance planning.

TABLE 2.8
Advanced PAT Implementation Framework

PAT Component	Implementation Level	Quality Impact	Regulatory Considerations
Spectroscopic analysis	In-line monitoring	Real-time quality assurance	Method validation required
Multivariate analysis	Process optimization	Enhanced process understanding	Model validation needed
Control aystems	Automated adjustment	Reduced variability	Change control documentation

Machine learning algorithms have proven particularly valuable in identifying complex relationships between process parameters and product quality attributes. These algorithms can analyze historical production data to identify patterns and correlations that inform process optimization strategies. The ability to predict product quality based on early-stage process parameters has enabled more proactive approaches to quality control and process adjustment.

The implementation of AI-driven control systems has enabled more sophisticated approaches to process optimization, where multiple process parameters are automatically adjusted to maintain optimal conditions. These systems can consider complex interactions between different process parameters and make adjustments that would be difficult to determine through traditional control approaches. The development of these advanced control systems represents a significant advancement in process control capabilities and has contributed to improved process consistency and product quality.

2.7.4 Continuous Manufacturing Integration

The adoption of continuous manufacturing principles in biomanufacturing represents a fundamental shift from traditional batch processing approaches. Continuous manufacturing offers several advantages, including improved process consistency, reduced production costs, and enhanced flexibility in production scheduling. The implementation of continuous manufacturing requires careful consideration of material flow, process control, and quality assurance strategies.

The transition to continuous manufacturing has been facilitated by advances in process analytical technology and control systems that enable real-time monitoring and adjustment of process parameters. These systems ensure consistent product quality while maintaining the efficiency advantages of continuous processing. The implementation of continuous manufacturing has also required the development of new approaches to material handling and inventory management, as traditional batch-based systems are not well suited to continuous processing operations.

The integration of continuous manufacturing principles has led to significant improvements in process efficiency and product quality. Continuous processing reduces the variability associated with batch-to-batch transitions and enables more consistent product quality. The reduced process complexity and improved efficiency of continuous manufacturing have made it an increasingly attractive option for biomanufacturing operations.

2.7.5 Sustainability and Green Manufacturing Initiatives

The implementation of sustainability initiatives in biomanufacturing has become increasingly important as organizations seek to reduce their environmental impact while maintaining operational efficiency. These initiatives encompass various aspects of manufacturing operations, including energy consumption, waste reduction, and material sourcing. The development of sustainable manufacturing practices requires careful consideration of environmental impacts throughout the product life cycle.

Modern biomanufacturing facilities incorporate various sustainability features, including energy-efficient equipment, waste recovery systems, and environmentally friendly materials. The implementation of these features has required significant investment in infrastructure and process modification but has resulted in reduced environmental impact and operating costs. The development of sustainable manufacturing practices has also led to improved public perception and regulatory compliance.

The integration of sustainability principles in biomanufacturing extends beyond facility design and operation to encompass supply chain management and material sourcing. Organizations are increasingly considering environmental impacts when selecting suppliers and materials, leading to the development of more sustainable supply chain networks. These considerations have influenced how organizations approach raw material management and process design.

2.7.6 Regulatory Compliance and Quality Management Evolution

The evolution of regulatory requirements and quality management systems in biomanufacturing continues to shape how organizations approach manufacturing operations. Modern quality management systems integrate traditional quality control approaches with advanced monitoring and control technologies to ensure consistent compliance with regulatory requirements. The implementation of these systems requires careful consideration of documentation, training, and validation requirements.

The increasing complexity of biomanufacturing processes and regulatory requirements has led to the development of more sophisticated quality management systems. These systems incorporate electronic documentation, automated compliance monitoring, and advanced risk management tools to ensure consistent regulatory compliance. The evolution of quality management systems has also been influenced by advances in technology and changing regulatory expectations.

The integration of quality management systems with manufacturing operations has become increasingly seamless, with quality considerations embedded throughout the manufacturing process. This integration has led to improved compliance outcomes and reduced quality-related issues. The development of more sophisticated quality management systems has also enabled more efficient approaches to regulatory compliance and quality assurance.

2.8 CONCLUSION

The comprehensive examination of raw material management and compliance in biomanufacturing presented in this chapter reveals the intricate interplay between traditional manufacturing principles and emerging technological innovations. The evolution of biomanufacturing processes has necessitated a fundamental shift in how organizations approach raw material management, quality control, and regulatory compliance. This transformation has been driven by the increasing complexity of biological products, heightened regulatory scrutiny, and the emergence of advanced technologies that enable more sophisticated approaches to process control and optimization.

The integration of raw material management strategies with compliance requirements has emerged as a critical factor in successful biomanufacturing operations. Organizations must maintain a delicate balance between operational efficiency and regulatory compliance, ensuring that all processes meet stringent quality standards while remaining economically viable. The implementation of comprehensive quality management systems, supported by robust documentation and training programs, has become essential for maintaining this balance. These systems must be sufficiently flexible to accommodate evolving regulatory requirements while providing the structure necessary for consistent product quality.

The adoption of lean manufacturing principles in biomanufacturing has demonstrated the potential for significant improvements in operational efficiency without compromising quality or compliance standards. The successful implementation of these principles requires careful consideration of the unique challenges presented by biological processes and regulatory requirements. Organizations that have effectively integrated lean principles with traditional quality management systems have achieved notable improvements in process efficiency, waste reduction, and overall operational performance (Figure 2.8).

The emergence of artificial intelligence and advanced analytics has revolutionized raw material management in biomanufacturing. These technologies have enabled more sophisticated approaches to quality prediction, process optimization, and supply chain management. The integration of AI-driven systems with traditional quality control methods has created new paradigms for ensuring consistent product quality while improving operational efficiency. The continued evolution of these technologies promises even greater advances in process control and optimization, potentially transforming how organizations approach raw material management and compliance.

The importance of sustainability in biomanufacturing operations has become increasingly apparent, influencing how organizations approach raw material management and process optimization. The development of sustainable manufacturing practices, including improved resource utilization and waste reduction strategies, has become a critical consideration in modern biomanufacturing operations.

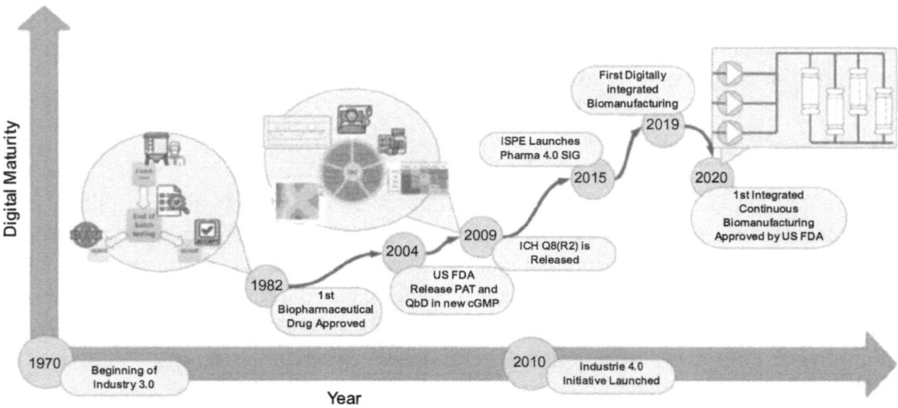

FIGURE 2.8 Integrated framework for future biomanufacturing excellence.

Organizations must balance these sustainability objectives with traditional goals of quality, efficiency, and compliance, creating new challenges and opportunities in raw material management.

Looking forward, the field of biomanufacturing continues to evolve rapidly, driven by technological innovations and changing regulatory requirements. The successful integration of advanced technologies with traditional manufacturing principles will be critical for organizations seeking to maintain competitive advantages while ensuring consistent product quality. The development of more sophisticated approaches to raw material management, supported by advanced analytics and artificial intelligence, will likely lead to further improvements in process efficiency and product quality.

The continued advancement of biomanufacturing technologies and processes will require ongoing adaptation of raw material management strategies and compliance frameworks. Organizations must remain vigilant in monitoring technological developments and regulatory changes, ensuring that their operations evolve to meet new challenges and opportunities. The successful integration of these various elements – technology, compliance, and sustainability – will be critical for organizations seeking to maintain leadership positions in the biomanufacturing industry.

In conclusion, the field of biomanufacturing stands at a critical juncture, where traditional manufacturing principles intersect with emerging technologies and evolving regulatory requirements. The successful navigation of these various elements requires a comprehensive understanding of raw material management principles, quality control requirements, and compliance frameworks. Organizations that can effectively integrate these elements while maintaining focus on operational efficiency and product quality will be best positioned for success in the evolving biomanufacturing landscape. The continued evolution of technology and regulatory requirements ensures that this field will remain dynamic, requiring ongoing adaptation and innovation from industry participants.

3 Process Analytical Technology

3.1 INTRODUCTION

Process analytical technology (PAT) stands as a transformative force in modern industrial processes, representing a fundamental paradigm shift in how manufacturing operations are conceived, implemented, and optimized. This revolutionary approach to process management and control has redefined the boundaries of manufacturing excellence, introducing unprecedented levels of process understanding and control capabilities that were previously unattainable. The comprehensive framework of PAT encompasses not just the technological tools and methodologies for process analysis and control but also embodies a philosophical transformation in how industries approach quality assurance and process optimization. This chapter delves deep into the multifaceted aspects of PAT, exploring its fundamental principles, technological implementations, and the profound impact it has had on manufacturing excellence across diverse industrial sectors.

The journey of process analytical technology from its conceptual origins to its current state of implementation represents a remarkable evolution in industrial process management. Traditional manufacturing processes, historically reliant on end-product testing and retrospective quality assurance measures, often struggled with significant challenges related to product consistency, waste reduction, and operational efficiency. These conventional approaches, while serving their purpose in earlier manufacturing paradigms, proved increasingly inadequate in meeting the demanding requirements of modern manufacturing operations. The emergence of PAT has fundamentally transformed this landscape, introducing a proactive and systematic approach to process understanding and control that addresses these historical limitations while opening new horizons for manufacturing excellence. This transformation has been particularly impactful in highly regulated industries such as pharmaceuticals, biotechnology, and advanced materials manufacturing, where product quality and process consistency are not just operational goals but also regulatory requirements.

The foundational principles of PAT extend far beyond simple process monitoring and control. At its core, PAT represents a comprehensive system for building quality into products through deep process understanding and real-time control mechanisms. This approach encompasses a wide array of technological elements, analytical methodologies, and control strategies that work in harmony to achieve optimal process performance. The systematic integration of advanced analytical instruments, sophisticated data processing algorithms, and intelligent control systems enables

DOI: 10.1201/9781003624592-3

continuous monitoring and adjustment of critical process parameters. This real-time capability ensures consistent product quality while simultaneously minimizing operational variability and waste. The holistic nature of PAT implementation creates a synergistic environment where process understanding, quality assurance, and operational efficiency converge to create sustainable manufacturing excellence (Figure 3.1).

The technological infrastructure supporting PAT implementation represents a complex ecosystem of interconnected systems and methodologies. Modern PAT frameworks incorporate multiple layers of functionality, ranging from fundamental process measurements to advanced predictive analytics and adaptive control strategies. This multilayered approach ensures robust process control while providing valuable insights into process behavior and product quality attributes. The analytical capabilities of PAT systems have evolved significantly, incorporating sophisticated spectroscopic methods, advanced sensor technologies, and real-time monitoring systems. These technological advancements have enabled unprecedented levels of process insight and control precision, fundamentally changing how manufacturers approach process optimization and quality assurance.

The implementation of PAT systems has catalyzed a fundamental shift in manufacturing culture, fostering an environment of continuous improvement and knowledge-based decision-making. This cultural transformation extends beyond the technical aspects of process control, encompassing organizational learning, knowledge management, and operational excellence. The systematic approach to process understanding and control enabled by PAT has created new paradigms for manufacturing excellence, where real-time process insight drives continuous optimization

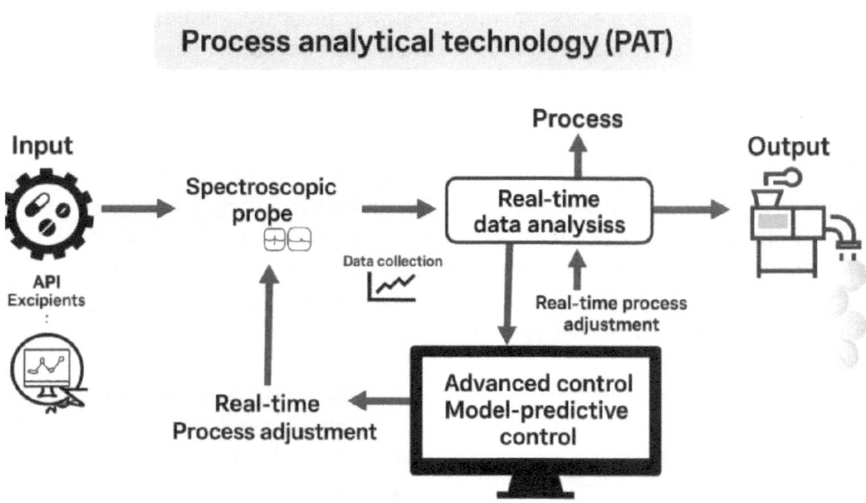

FIGURE 3.1 Integrated PAT framework architecture.

and quality improvement. This comprehensive approach to process management has proven particularly valuable in regulated industries, where product quality and process consistency are critical requirements that must be maintained consistently across production batches and operational cycles.

The evolution of PAT continues to be driven by rapid advancements in technology and manufacturing methodologies. The integration of artificial intelligence (AI) and machine learning algorithms within the PAT framework represents a new frontier in process optimization and control. These advanced technologies enable sophisticated predictive analytics, pattern recognition capabilities, and adaptive control strategies that further enhance process understanding and operational excellence. The synergy between traditional PAT methodologies and emerging AI technologies creates powerful new capabilities for process optimization and quality assurance, pushing the boundaries of what is possible in manufacturing excellence.

The digital transformation of manufacturing operations has positioned PAT as a cornerstone of Industry 4.0 initiatives. The comprehensive framework provided by PAT aligns perfectly with the principles of smart manufacturing, enabling the development of intelligent production systems that can adapt and optimize in real time. This evolution toward smart manufacturing represents a natural progression of PAT principles, leveraging advanced technologies to achieve unprecedented levels of process control and product quality. The integration of PAT within the broader context of digital manufacturing creates new opportunities for process optimization, quality assurance, and operational excellence.

The significance of PAT extends well beyond its immediate impact on process control and quality assurance. The framework provides a foundation for sustainable manufacturing excellence, enabling organizations to optimize their operations while minimizing environmental impact and resource consumption. The real-time monitoring and control capabilities of PAT systems facilitate more efficient resource utilization, reduced waste generation, and improved energy efficiency. These sustainability benefits, coupled with the operational advantages of PAT implementation, create a compelling case for its adoption across diverse industrial sectors.

The subsequent sections of this chapter will explore in detail the various components and methodologies that constitute modern PAT systems. From data acquisition tools and process analyzers to control strategies and knowledge management systems, each element plays a crucial role in the overall effectiveness of PAT implementation. Special attention will be given to emerging technologies and methodologies that are shaping the future of process analytical technology, providing readers with a comprehensive understanding of both current capabilities and future directions in this rapidly evolving field.

3.2 BACKGROUND AND NEED FOR PAT

3.2.1 Historical Evolution of Process Analysis

The evolution of process analysis in industrial manufacturing represents a remarkable journey spanning several decades of technological advancement and philosophical transformation in quality management approaches. In the early stages of industrial manufacturing, process analysis was predominantly based on empirical observations and end-product testing methodologies that provided limited insight into the actual manufacturing process. This retrospective approach to quality assurance, while standard practice for many years, presented significant limitations in terms of process understanding and control capabilities. The manufacturing industry's reliance on these traditional methods often resulted in substantial product wastage, quality variations, and operational inefficiencies that impacted both productivity and profitability. The recognition of these limitations, coupled with increasing regulatory requirements and market demands for consistent product quality, created a compelling need for more sophisticated approaches to process analysis and control.

The transition from traditional quality control methods to modern process analytical technologies was catalyzed by several factors that emerged during the latter half of the 20th century. The increasing complexity of manufacturing processes, particularly in industries such as pharmaceuticals and biotechnology, demanded more sophisticated approaches to process understanding and control. Traditional end-product testing methodologies proved inadequate in addressing the challenges posed by these complex processes, leading to the development of more advanced analytical techniques and monitoring systems. The emergence of new analytical technologies, coupled with advances in computer systems and data processing capabilities, created opportunities for real-time process monitoring and control that were previously unattainable (Figure 3.2).

3.2.2 Fundamental Drivers for PAT Implementation

The fundamental drivers for PAT implementation extend far beyond simple process optimization and quality control considerations. The increasing globalization of manufacturing operations has created new challenges in maintaining consistent product quality across different production facilities and geographical locations. This globalization effect has amplified the need for standardized approaches to process analysis and control that can ensure consistent product quality regardless of manufacturing location. The implementation of PAT systems provides a framework for achieving this standardization while simultaneously enabling local optimization of process parameters based on specific operational conditions and requirements.

The regulatory landscape has played a pivotal role in driving PAT adoption across various industries. Regulatory agencies worldwide have recognized the value of PAT in ensuring product quality and process consistency, leading to the development of guidelines and requirements that encourage its implementation. The FDA's PAT initiative, launched in the early 2000s, represents a significant milestone in this regulatory evolution, establishing a framework for implementing modern process analytical technologies in regulated industries. This regulatory support has been instrumental in driving widespread adoption of PAT across various industrial sectors, particularly in pharmaceutical and biotechnology manufacturing.

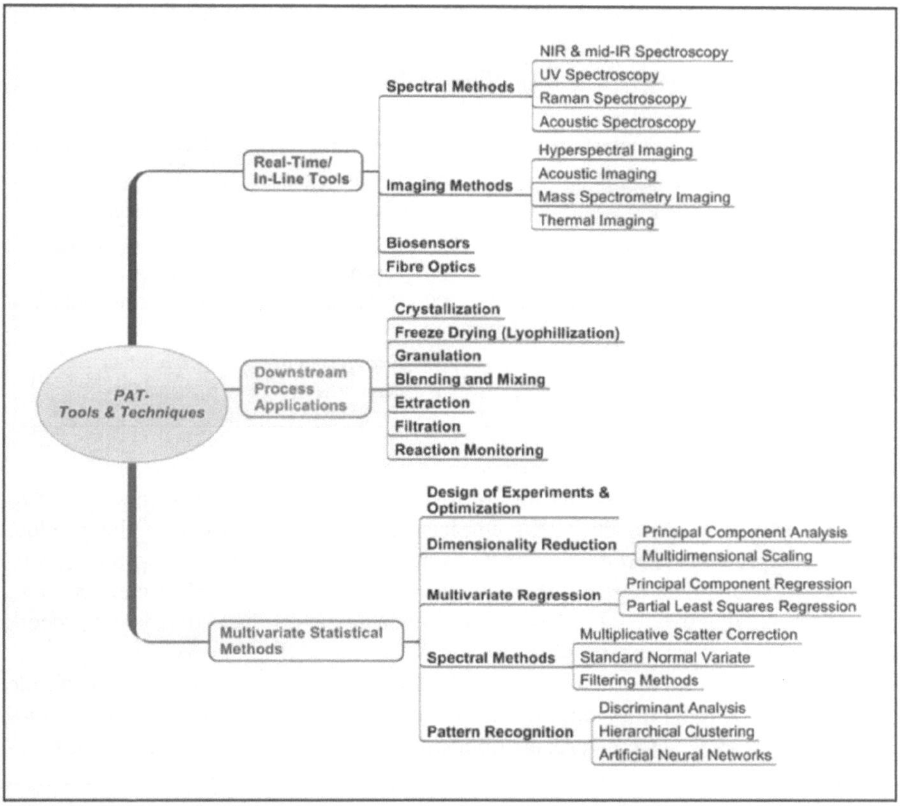

FIGURE 3.2 Evolution timeline of process analysis: From traditional methods to modern PAT implementation.

3.2.3 Economic and Operational Benefits

The economic implications of PAT implementation represent a compelling business case for organizations considering its adoption. The initial investment in PAT infrastructure, while significant, is often justified by the substantial long-term benefits in terms of reduced operational costs, improved product quality, and enhanced manufacturing efficiency. The real-time monitoring and control capabilities enabled by PAT systems result in significant reductions in product waste, optimization of raw material usage, and improvements in overall equipment effectiveness. These operational improvements translate directly into tangible economic benefits that contribute to the organization's bottom line performance (Table 3.1).

3.2.4 Quality Assurance Paradigm Shift

The implementation of PAT has fundamentally transformed the approach to quality assurance in manufacturing operations. Traditional quality control methods, based

TABLE 3.1
Economic Impact Analysis of PAT Implementation

Benefit Category	Impact Areas	Quantitative Benefits	Qualitative Benefits
Cost reduction	Manufacturing operations	15%–25% reduction in operating costs	Enhanced process understanding
Quality improvement	Product consistency	30%–40% reduction in quality deviations	Improved regulatory compliance
Efficiency gains	Production throughput	20%–30% increase in production efficiency	Better resource utilization
Waste reduction	Material usage	25%–35% reduction in material waste	Environmental sustainability

primarily on end-product testing and retrospective analysis, have been superseded by more proactive approaches that emphasize building quality into the product through comprehensive process understanding and control. This paradigm shift represents a fundamental change in how organizations approach quality management, moving from a reactive to a proactive stance in ensuring product quality and process consistency.

The quality assurance benefits of PAT implementation extend beyond simple compliance with regulatory requirements. The comprehensive process understanding enabled by PAT systems creates opportunities for continuous quality improvement through real-time monitoring and control of critical process parameters. This approach to quality assurance enables organizations to identify and address potential quality issues before they impact the final product, resulting in significant improvements in product consistency and overall quality levels.

3.2.5 Technical Infrastructure Requirements

The implementation of PAT systems necessitates a robust technical infrastructure that can support the various components and functionalities required for effective process analysis and control. This infrastructure encompasses multiple elements, including analytical instruments, data acquisition systems, control interfaces, and information management platforms. The integration of these various components requires careful consideration of system architecture, communication protocols, and data management strategies to ensure optimal performance and reliability.

The technical requirements for PAT implementation vary significantly based on the specific application context and industry requirements. However, certain fundamental elements remain consistent across different implementations. These include robust analytical instrumentation capabilities, reliable data acquisition and processing systems, and effective control interfaces that enable real-time process adjustments based on analytical results. The selection and integration of these various components require careful consideration of both technical capabilities and operational requirements to ensure optimal system performance.

3.2.6 Organizational Impact and Change Management

The successful implementation of PAT systems requires significant organizational changes that extend beyond technical considerations. The adoption of PAT methodologies often necessitates changes in operational procedures, workforce skills, and organizational culture. These changes must be carefully managed to ensure successful implementation and sustainable operation of PAT systems. The development of appropriate training programs, standard operating procedures, and change management strategies plays a crucial role in ensuring successful PAT implementation.

The organizational impact of PAT implementation extends to various functional areas within the organization. From operational staff responsible for day-to-day process monitoring and control to quality assurance personnel involved in compliance monitoring and documentation, the implementation of PAT systems affects multiple stakeholders across the organization. Successfully managing these organizational impacts requires a comprehensive change management approach that addresses both technical and human factors involved in PAT implementation.

3.3 TOOLS FOR DATA ACQUISITION

3.3.1 Overview of Data Acquisition Systems

The foundation of effective process analytical technology implementation lies in robust and reliable data acquisition systems that form the cornerstone of process understanding and control. These sophisticated systems encompass a wide array of technologies, methodologies, and instruments designed to capture critical process parameters with unprecedented accuracy and reliability. Modern data acquisition systems have evolved significantly from their early predecessors, incorporating advanced sensing technologies, high-speed data processing capabilities, and intelligent analysis algorithms that enable real-time process monitoring and control. The comprehensive nature of these systems ensures that manufacturers can maintain consistent product quality while optimizing operational efficiency across their production processes.

Data acquisition in PAT systems represents a complex interplay of hardware components, software systems, and analytical methodologies working in concert to provide comprehensive process insight. The evolution of these systems has been driven by advances in sensor technology, computing capabilities, and analytical methodologies, resulting in increasingly sophisticated platforms capable of capturing and processing vast amounts of process data in real time. This technological progression has enabled manufacturers to implement more effective control strategies, optimize their processes with greater precision, and ensure consistent product quality across production batches.

The architecture of modern data acquisition systems reflects the complex requirements of contemporary manufacturing operations. These systems typically incorporate multiple layers of functionality, from basic sensor networks to sophisticated data processing and analysis capabilities. The hierarchical structure ensures efficient data flow while maintaining system reliability and data integrity. This layered approach enables manufacturers to implement comprehensive monitoring strategies that capture both macro-level process trends and micro-level quality attributes, providing a complete picture of process performance and product quality.

The integration of advanced sensing technologies within PAT data acquisition systems has revolutionized process monitoring capabilities. Modern systems incorporate a diverse array of sensor types, including spectroscopic analyzers, chromatographic systems, and advanced physical sensors, each selected to monitor specific process parameters or quality attributes. This multimodal approach to process monitoring ensures comprehensive coverage of critical process parameters while providing redundancy and verification capabilities that enhance system reliability.

3.3.2 Advanced Sensing Technologies

The advancement of sensing technologies has played a pivotal role in enhancing the capabilities of PAT data acquisition systems. Modern manufacturing operations rely on a diverse array of sensor types, each designed to capture specific aspects of process performance and product quality. Spectroscopic methods, including near-infrared (NIR), Raman spectroscopy, and UV-visible spectrophotometry, have become particularly important in process monitoring applications. These technologies enable real-time analysis of chemical composition, physical properties, and quality attributes without the need for sample preparation or destructive testing.

Physical sensors form another critical component of PAT data acquisition systems, providing continuous monitoring of parameters such as temperature, pressure, flow rates, and viscosity. The integration of these sensors within the PAT framework enables comprehensive process monitoring while ensuring that all critical parameters remain within specified control limits. Advanced sensor designs incorporate features such as self-diagnostics, automatic calibration capabilities, and intelligent signal processing algorithms that enhance measurement reliability and reduce maintenance requirements.

The evolution of microelectromechanical systems (MEMS) technology has enabled the development of increasingly sophisticated sensor platforms that combine multiple measurement capabilities within compact, reliable packages. These integrated sensor systems provide comprehensive process monitoring capabilities while minimizing installation complexity and maintenance requirements. The miniaturization of sensor components has also enabled the implementation of monitoring capabilities in previously inaccessible process locations, enhancing the overall coverage and effectiveness of PAT systems.

3.3.3 Data Processing and Signal Analysis

The processing of raw sensor data represents a critical aspect of PAT data acquisition systems. Modern systems incorporate sophisticated signal processing algorithms that convert raw sensor outputs into meaningful process information while filtering out noise and accounting for measurement uncertainties. These processing capabilities ensure that the data provided to control systems and operators accurately reflects actual process conditions, enabling effective decision-making and control actions.

Signal analysis methodologies in PAT systems have evolved to incorporate advanced mathematical techniques that enhance data quality and reliability. Multivariate analysis methods, pattern recognition algorithms, and statistical processing techniques work together to extract meaningful information from complex sensor signals. These analytical capabilities enable the identification of subtle process trends and potential quality issues that might not be apparent from individual sensor measurements.

The implementation of real-time data processing capabilities has significantly enhanced the effectiveness of PAT systems. Modern platforms can process vast amounts of sensor data in real time, enabling immediate response to process variations or quality issues. This real-time processing capability is particularly important in continuous manufacturing operations, where rapid detection and correction of process deviations are essential for maintaining product quality and operational efficiency.

3.3.4 Data Management and Storage Systems

The management and storage of process data represent critical aspects of PAT implementation. Modern systems incorporate sophisticated data management platforms that ensure secure storage while enabling efficient access to historical process information. These systems must balance the need for comprehensive data retention with practical considerations related to storage capacity and access speed. The implementation of effective data management strategies ensures that valuable process information remains available for analysis and optimization activities while maintaining system performance and reliability (Table 3.2).

TABLE 3.2
Components of PAT Data Management Systems

System Component	Primary Function	Operational Impact	Integration Level
Data historians	Long-term storage	Process trending	Enterprise-wide
Real-time databases	Immediate access	Control systems	Unit operation
Analytics platforms	Data analysis	Process optimization	Plant-wide
Knowledge bases	**Information management**	Decision support	Organization

3.3.5 System Integration and Communication

The integration of data acquisition systems within the broader manufacturing infrastructure requires careful consideration of communication protocols and interface requirements. Modern PAT systems must communicate effectively with various plant systems, including control platforms, manufacturing execution systems (MESs), and enterprise resource planning (ERP) systems. The implementation of standardized communication protocols ensures seamless data flow while maintaining system security and reliability.

Network architecture plays a crucial role in PAT system implementation, providing the infrastructure necessary for reliable data transmission and system communication. Modern systems typically incorporate redundant network paths, advanced security features, and intelligent routing capabilities that ensure reliable operation under various conditions. The design of network infrastructure must consider factors such as data volume, transmission speed requirements, and system reliability requirements to ensure effective operation of the PAT system.

3.3.6 Quality Assurance and Validation

The validation of data acquisition systems represents a critical aspect of PAT implementation, particularly in regulated industries. Comprehensive validation protocols ensure that systems operate reliably and produce accurate results under all operating conditions. The implementation of effective validation strategies requires careful consideration of regulatory requirements, operational needs, and system capabilities to ensure compliance while maintaining operational efficiency.

The maintenance of data quality requires ongoing attention to system calibration, maintenance procedures, and performance verification activities. Modern PAT systems incorporate automated calibration capabilities and performance monitoring features that help ensure continued accuracy and reliability. The implementation of comprehensive quality assurance programs ensures that systems maintain their performance capabilities while meeting regulatory requirements and operational needs.

3.4 SOFTWARE IN FERMENTERS, FLOW FILTRATIONS, CHROMATOGRAPHY

3.4.1 Integrated Software Architecture for Bioprocess Systems

The implementation of software systems in modern bioprocessing operations represents a complex interplay of multiple technological frameworks and control methodologies. These sophisticated software architectures form the backbone of contemporary fermentation, filtration, and chromatographic processes, enabling unprecedented levels of process control and optimization. The integration of these software systems extends far beyond simple process monitoring, encompassing comprehensive data management, advanced process control, and sophisticated analytical capabilities that work in concert to ensure optimal process performance and product quality. Modern bioprocess software architectures are designed with modularity and scalability in mind, allowing for seamless integration of various process units while maintaining the flexibility to accommodate future technological advancements and changing operational requirements.

The fundamental architecture of bioprocess software systems incorporates multiple layers of functionality, each serving specific operational requirements while contributing to the overall system effectiveness. At the core of these systems lies a robust data management infrastructure that facilitates the collection, processing, and storage of process data from various sources including sensors, analyzers, and control systems. This data management layer serves as the foundation for higher-level functions including process control, optimization, and quality assurance. The integration of these various software components creates a comprehensive system that enables real-time process monitoring, control, and optimization while maintaining compliance with regulatory requirements and quality standards (Figure 3.3).

The evolution of bioprocess software systems has been driven by advances in computing technology, data management capabilities, and process control methodologies. Modern systems incorporate sophisticated algorithms for process optimization,

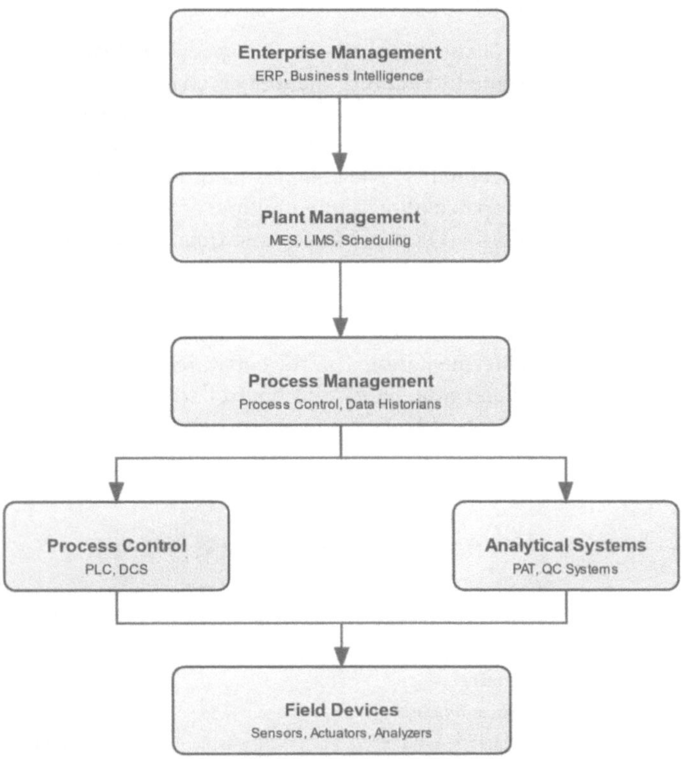

FIGURE 3.3 Hierarchical architecture of integrated bioprocess software systems.

predictive maintenance, and quality assurance, enabling a level of process control and optimization that was previously unattainable. The integration of artificial intelligence and machine learning capabilities has further enhanced these systems, enabling predictive analytics and adaptive control strategies that can optimize process performance in real time. These technological advancements have transformed bioprocess operations, enabling unprecedented levels of process understanding and control while maintaining strict compliance with regulatory requirements and quality standards.

The implementation of bioprocess software systems requires careful consideration of various operational requirements, regulatory compliance needs, and system integration challenges. Modern systems must be designed to handle large volumes of process data while maintaining system performance and data integrity. The software architecture must support various data types and formats, from simple sensor readings to complex analytical data from sophisticated process analyzers. The system must also provide robust data security and audit trail capabilities to ensure compliance with regulatory requirements and maintain data integrity throughout the process life cycle.

3.4.2 Fermentation Process Control Software

Fermentation process control software represents a specialized category of bioprocess software systems designed to address the unique challenges and requirements of fermentation operations. These sophisticated software systems integrate multiple control strategies and analytical capabilities to optimize fermentation performance and ensure consistent product quality. The complex nature of fermentation processes requires software systems capable of handling multiple process variables simultaneously while maintaining optimal conditions for microbial growth and product formation. Modern fermentation control software incorporates advanced algorithms for process optimization, including feed rate control, dissolved oxygen management, and metabolic flux analysis (Table 3.3).

The implementation of fermentation control software requires careful consideration of various operational parameters and control strategies. Modern systems

TABLE 3.3
Core Components of Modern Fermentation Control Software

Component Category	Primary Functions	Integration Level	Key Features
Process control	Fermentation parameters management	Unit operation	Real-time control
Data analytics	Performance monitoring	Process-wide	Predictive analytics
Quality management	Product quality assurance	Facility-wide	Compliance tools
System integration	Equipment communication	Enterprise	Standardized protocols

incorporate sophisticated algorithms for monitoring and controlling critical process parameters including temperature, pH, dissolved oxygen, and nutrient concentrations. These control strategies must be capable of handling the dynamic nature of fermentation processes while maintaining optimal conditions for product formation. The software must also provide robust data management capabilities to ensure accurate tracking of process parameters and maintain compliance with regulatory requirements.

The evolution of fermentation control software has been marked by significant advancements in control strategies and analytical capabilities. Modern systems incorporate sophisticated modeling capabilities that enable predictive control strategies and process optimization. These advanced control capabilities are complemented by robust data management and analysis tools that provide comprehensive insight into process performance and product quality. The integration of artificial intelligence and machine learning algorithms has further enhanced these capabilities, enabling adaptive control strategies that can optimize process performance in real time.

3.4.3 Flow Filtration Software Systems

Flow filtration software systems represent another critical component of modern bioprocess operations, enabling precise control and optimization of filtration processes. These specialized software systems are designed to handle the unique challenges associated with various filtration operations, from simple dead-end filtration to complex tangential flow filtration processes. The software architecture must support various filtration modes while providing robust control capabilities and comprehensive data management functions. Modern filtration software systems incorporate advanced algorithms for optimizing filtration parameters, monitoring filter performance, and ensuring consistent product quality.

The implementation of flow filtration software requires careful consideration of various operational parameters and control strategies. Modern systems incorporate sophisticated algorithms for monitoring and controlling critical filtration parameters including transmembrane pressure, flux rates, and filter integrity. These control strategies must be capable of handling various filtration modes while maintaining optimal performance and product quality. The software must also provide robust data management capabilities to ensure accurate tracking of filtration parameters and maintain compliance with regulatory requirements.

The complexity of modern filtration operations necessitates software systems capable of handling multiple process variables simultaneously while maintaining optimal filtration performance. These systems must provide comprehensive monitoring and control capabilities while ensuring data integrity and regulatory compliance. The integration of advanced analytical capabilities enables real-time optimization of filtration parameters, ensuring consistent product quality while maximizing process efficiency. The software architecture must support various filtration modes while providing flexible control strategies that can adapt to changing process conditions.

3.4.4 Chromatography Software Integration

Chromatography software systems represent some of the most sophisticated and complex software implementations in modern bioprocess operations. These specialized systems must handle the intricate requirements of various chromatographic processes while providing robust control capabilities and comprehensive data analysis functions. Modern chromatography software systems integrate multiple analytical capabilities and control strategies to optimize separation performance and ensure consistent product quality. The software architecture must support various chromatographic modes while providing flexible control strategies that can adapt to changing process conditions.

The implementation of chromatography software requires careful consideration of various operational parameters and control strategies. Modern systems incorporate sophisticated algorithms for monitoring and controlling critical chromatographic parameters including flow rates, buffer composition, and column performance. These control strategies must be capable of handling various separation modes while maintaining optimal performance and product quality. The software must also provide robust data management capabilities to ensure accurate tracking of chromatographic parameters and maintain compliance with regulatory requirements.

The evolution of chromatography software has been driven by advances in separation technology and analytical capabilities. Modern systems incorporate sophisticated modeling capabilities that enable predictive control strategies and process optimization. These advanced control capabilities are complemented by robust data management and analysis tools that provide comprehensive insight into separation performance and product quality. The integration of artificial intelligence and machine learning algorithms has further enhanced these capabilities, enabling adaptive control strategies that can optimize separation performance in real time.

3.5 ANALYSIS AND DESIGN PROCESS ANALYZERS

3.5.1 Fundamentals of Process Analyzer Design

The design and implementation of process analyzers represents a critical component within the broader framework of process analytical technology (PAT). These sophisticated analytical instruments serve as the primary interfaces between physical processes and control systems, providing essential data for real-time monitoring and process optimization. Process analyzer design encompasses a complex interplay of analytical chemistry, engineering principles, and system integration considerations, requiring a thorough understanding of both theoretical foundations and practical implementation challenges. The development of effective process analyzers demands careful consideration of multiple factors, including sample handling requirements, analytical methodology selection, instrument robustness, and integration capabilities with existing process control systems.

Modern process analyzer design has evolved significantly from its early origins, incorporating advanced materials, sophisticated electronics, and intelligent data processing capabilities. The fundamental architecture of process analyzers typically includes multiple subsystems working in concert to provide accurate and reliable analytical data. These subsystems encompass sample acquisition and conditioning components, analytical measurement systems, data processing units, and communication interfaces. The integration of these elements requires careful consideration of various operational parameters, including response time requirements, measurement accuracy specifications, and maintenance accessibility. The design process must also account for environmental factors such as temperature variations, vibration effects, and potential chemical exposure, ensuring reliable operation under actual process conditions.

Process analyzers in contemporary manufacturing environments must demonstrate exceptional reliability and accuracy while maintaining operational efficiency. The design philosophy emphasizes the importance of robust analytical performance coupled with practical considerations such as ease of maintenance, calibration stability, and system diagnostics. Modern analyzer designs increasingly incorporate self-diagnostic capabilities, automated calibration features, and predictive maintenance functionalities, enabling proactive system management and minimal operational disruption. These advanced features contribute significantly to the overall effectiveness of process analytical systems, ensuring consistent performance and reducing the need for frequent manual intervention (Figure 3.4).

The selection and implementation of appropriate analytical methodologies represents a crucial aspect of process analyzer design. Contemporary analyzers employ various analytical techniques, including spectroscopic methods, chromatographic separations, and electrochemical measurements. The choice of analytical methodology must consider factors such as measurement speed requirements, sample matrix complexity, and desired analytical precision. Additionally, the selected methodology

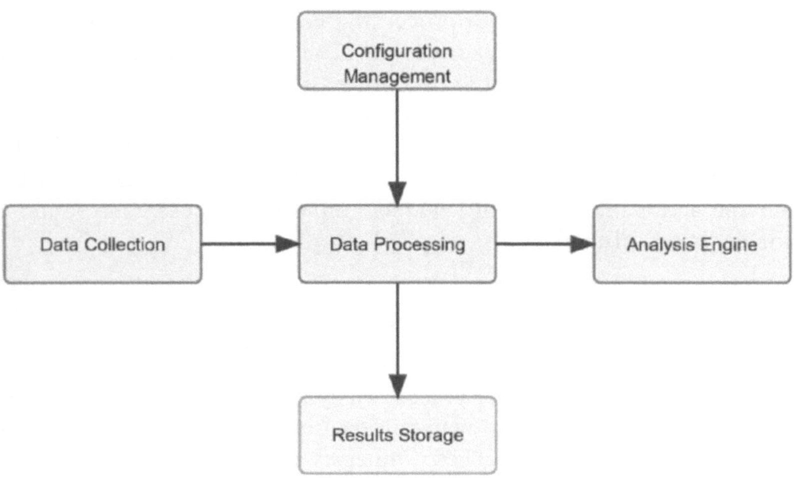

FIGURE 3.4 Integrated process analyzer architecture.

must be compatible with the process environment and capable of providing reliable results under actual operating conditions. The integration of multiple analytical techniques within a single analyzer platform often provides complementary information, enabling more comprehensive process understanding and control.

3.5.2 Sample Handling and Conditioning Systems

The development of effective sample handling and conditioning systems represents a fundamental challenge in process analyzer design. These systems must ensure that representative samples are consistently delivered to the analytical instrument while maintaining sample integrity and measurement reliability. The design of sample handling systems requires careful consideration of factors such as sample flow rates, pressure requirements, and temperature control needs. The implementation of appropriate sample conditioning techniques, including filtration, pressure regulation, and temperature adjustment, ensures that samples arrive at the analyzer in a suitable state for accurate analysis.

Sample system design must address various practical considerations, including materials compatibility, maintenance accessibility, and system reliability. The selection of appropriate materials for sample lines, valves, and other components must account for chemical compatibility, temperature requirements, and mechanical stress factors. The implementation of proper safety measures, including pressure relief systems and containment features, ensures safe operation under all process conditions. Regular maintenance requirements must be considered during the design phase, incorporating features that facilitate routine cleaning, component replacement, and system verification.

3.5.3 Data Processing and Integration

The integration of advanced data processing capabilities represents a crucial aspect of modern process analyzer design. Contemporary analyzers incorporate sophisticated algorithms for signal processing, data validation, and result calculation. These processing capabilities enable real-time data analysis, automatic outlier detection, and intelligent system diagnostics. The implementation of appropriate data processing strategies ensures that analytical results are both accurate and actionable, providing valuable input for process control and optimization decisions (Table 3.4).

3.5.4 Calibration and Validation Strategies

The development and implementation of effective calibration and validation strategies represents a critical aspect of process analyzer operation. Modern analyzers incorporate various approaches to maintain measurement accuracy and reliability over extended operating periods. These strategies may include automated calibration systems, reference standard monitoring, and drift compensation techniques. The implementation of appropriate validation procedures ensures that analytical results

TABLE 3.4
Key Components of Process Analyzer Data Integration Systems

Component	Primary Function	Integration Level	Data Processing Capability
Signal processing	Raw data analysis	Analyzer level	Real-time processing
Data validation	Quality assurance	System level	Automated verification
Result calculation	Information generation	Enterprise level	Complex analytics
System diagnostics	Performance monitoring	Multiple levels	Predictive analysis

remain reliable and traceable over time, meeting both operational requirements and regulatory standards.

The design of calibration systems must consider practical aspects such as calibration frequency requirements, standard material stability, and system automation capabilities. The implementation of intelligent calibration scheduling, based on both time intervals and performance metrics, optimizes system reliability while minimizing operational disruption. The incorporation of appropriate validation procedures, including system suitability tests and performance verification protocols, ensures consistent analyzer performance and result reliability.

3.5.5 Maintenance and Reliability Considerations

The design of process analyzers must incorporate comprehensive maintenance and reliability considerations to ensure long-term operational success. These considerations encompass various aspects, including component accessibility, spare parts management, and preventive maintenance requirements. The implementation of appropriate maintenance strategies, including condition-based monitoring and predictive maintenance approaches, optimizes system reliability while minimizing operational disruption. The development of effective maintenance procedures and documentation ensures consistent analyzer performance and reliable operation over extended periods.

Reliability engineering principles play a crucial role in modern analyzer design, incorporating redundancy features, failure mode analysis, and risk mitigation strategies. The implementation of appropriate reliability measures, including component redundancy and fault-tolerant design approaches, ensures continuous system availability and minimal process disruption. The development of comprehensive maintenance documentation and training programs supports effective system management and optimal operational performance.

3.5.6 Future Trends and Development Directions

The evolution of process analyzer design continues to be driven by technological advances and changing industrial requirements. Emerging trends include the integration of artificial intelligence capabilities, enhanced miniaturization, and improved wireless communication features. These developments promise to further enhance analyzer performance, reliability, and integration capabilities. The incorporation of advanced materials and manufacturing techniques enables the development of more compact, efficient, and reliable analyzer systems.

The future of process analyzer design points toward increased intelligence and autonomy, with systems capable of self-diagnosis, adaptive calibration, and predictive maintenance. The integration of advanced data analytics and machine learning capabilities will enable more sophisticated data interpretation and process understanding. These developments will continue to enhance the role of process analyzers in supporting manufacturing excellence and operational optimization.

3.6 PROCESS CONTROL TOOLS

3.6.1 Advanced Control System Architecture

Process control tools represent the cornerstone of modern manufacturing systems, serving as the critical interface between process understanding and operational excellence. The architecture of advanced control systems in process analytical technology (PAT) encompasses a sophisticated network of interconnected components, each contributing to the overall goal of process optimization and quality assurance. These systems integrate multiple layers of control functionality, from basic regulatory control to advanced supervisory systems, creating a comprehensive framework for process management and optimization. The implementation of these control tools requires careful consideration of various factors, including process dynamics, measurement capabilities, and control objectives, to ensure optimal performance and reliability.

Modern control system architecture in PAT applications extends beyond traditional proportional-integral-derivative (PID) control schemes, incorporating advanced control strategies that can handle complex process interactions and constraints. These sophisticated control systems utilize model-based approaches, adaptive control algorithms, and intelligent decision-making capabilities to maintain optimal process conditions while responding to changing operational requirements. The integration of multiple control layers creates a hierarchical structure that enables both local optimization of individual process units and global optimization of the entire manufacturing system. This comprehensive approach to process control ensures robust performance while maintaining the flexibility needed to adapt to changing process conditions and production requirements.

The implementation of advanced control architectures necessitates careful consideration of system integration and communication protocols. Modern PAT systems utilize standardized communication interfaces and protocols to ensure seamless integration between different control components and analytical instruments. This integration enables real-time data exchange and coordination between various system elements, facilitating comprehensive process monitoring and control. The communication infrastructure must support both high-speed data transfer for real-time control applications and reliable data storage for historical analysis and regulatory compliance (Figure 3.5).

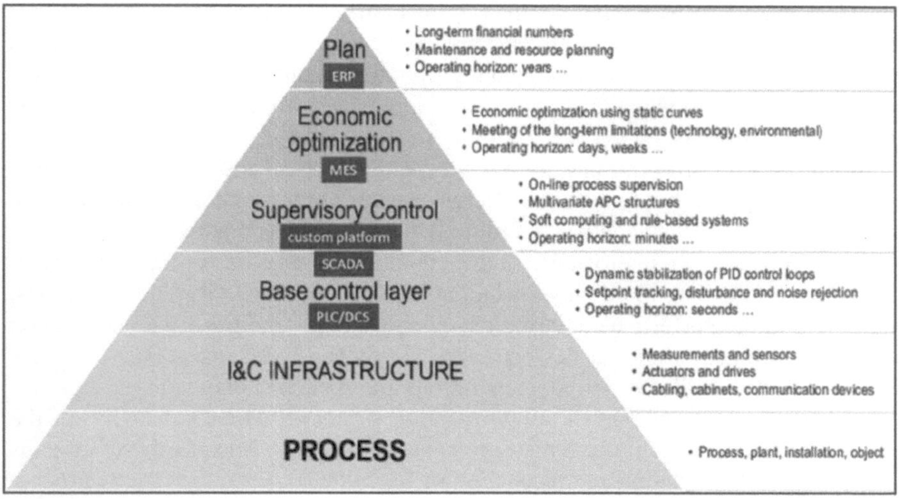

FIGURE 3.5 Hierarchical structure of advanced process control systems in PAT implementation.

The human-machine interface (HMI) plays a crucial role in the effectiveness of process control tools, serving as the primary point of interaction between operators and the control system. Modern HMI designs incorporate ergonomic principles and intuitive visualization techniques to present complex process information in an easily understandable format. The interface must support both routine operational tasks and emergency response scenarios, providing operators with clear information and straightforward control capabilities. The design of effective HMIs requires careful consideration of human factors engineering principles to ensure optimal operator performance and system reliability.

The reliability and robustness of control system architecture represent critical considerations in PAT implementations. Redundancy mechanisms, fault detection systems, and backup control strategies must be incorporated to ensure continuous operation even in the event of component failures or communication disruptions. The system architecture must also support regular maintenance and upgrades without compromising process control integrity. This requires careful planning of system components and communication pathways to enable modular replacement and enhancement of control system elements.

3.6.2 Real-Time Process Optimization Strategies

Real-time process optimization represents a fundamental capability of modern PAT systems, enabling continuous adjustment of process parameters to maintain optimal operating conditions. These optimization strategies utilize advanced algorithms and models to evaluate process performance and identify opportunities for improvement in real time. The implementation of real-time optimization requires careful consideration of process dynamics, measurement uncertainties, and operational constraints to ensure reliable and effective control actions. The optimization strategies must balance multiple objectives, including product quality, operational efficiency, and resource utilization while maintaining compliance with regulatory requirements and safety constraints.

Advanced optimization algorithms incorporate both steady-state and dynamic optimization approaches to address different aspects of process control. Steady-state optimization focuses on identifying optimal operating points that maximize process efficiency and product quality under stable conditions. Dynamic optimization strategies address transient operations, including batch processes and grade transitions, ensuring optimal performance during changing process conditions. The integration of these optimization approaches creates a comprehensive framework for process control that can handle both routine operations and special situations.

The effectiveness of real-time optimization strategies depends heavily on the quality and reliability of process measurements and models. Advanced PAT systems incorporate data validation and reconciliation techniques to ensure accurate process information for optimization calculations. Model adaptation and update mechanisms maintain the accuracy of process models used in optimization calculations, ensuring reliable performance even as process conditions change over time. The optimization system must also handle measurement uncertainties and process disturbances effectively to maintain robust performance under varying operational conditions (Table 3.5).

The implementation of real-time optimization strategies requires careful consideration of computational requirements and system response times. The optimization algorithms must execute quickly enough to provide timely control actions while maintaining solution quality and reliability. This often necessitates the use of specialized computational techniques and hardware configurations to achieve the required performance levels. The optimization system must also handle multiple time scales effectively, from rapid control actions to longer-term optimization objectives.

TABLE 3.5
Key Components of Real-Time Process Optimization Systems

Component	Primary Function	Implementation Considerations	Performance Metrics
Optimization algorithms	Process parameter adjustment	Computational efficiency	Solution quality
Process models	System behavior prediction	Model accuracy	Prediction error
Constraint handlers	Operating limits management	Safety compliance	Constraint satisfaction
Performance monitors	System evaluation	Real-time capability	Response time

The integration of real-time optimization with other PAT system components represents a critical aspect of system implementation. The optimization system must coordinate effectively with regulatory control systems, supervisory control layers, and data management systems to ensure coherent operation. This integration enables comprehensive process optimization while maintaining operational stability and regulatory compliance. The system must also provide appropriate feedback and documentation capabilities to support continuous improvement and regulatory requirements.

3.6.3 Advanced Feedback Control Mechanisms

Advanced feedback control mechanisms in PAT systems represent a sophisticated evolution of traditional control methodologies, incorporating multiple layers of feedback and adaptation to achieve superior process performance. These advanced mechanisms extend beyond conventional single-input-single-output control loops to encompass multivariable control strategies that can handle complex process interactions and dependencies. The implementation of these advanced control mechanisms requires careful consideration of process dynamics, measurement limitations, and control objectives to ensure optimal performance under varying operational conditions.

Modern feedback control systems in PAT applications utilize sophisticated algorithms that can adapt to changing process conditions while maintaining robust performance. These adaptive control mechanisms continuously monitor process behavior and adjust control parameters to optimize system response and stability. The integration of multiple feedback loops creates a comprehensive control structure that can handle both local disturbances and global process variations effectively. This multilayered approach to feedback control ensures consistent product quality while minimizing process variability and operational disruptions.

The development of effective feedback control strategies requires sophisticated modeling and analysis techniques to understand process behavior and design appropriate control responses. Advanced system identification methods enable the development of accurate process models that capture both steady-state and dynamic behavior. These models serve as the foundation for control system design, enabling the development of control strategies that can handle complex process interactions while maintaining stable operation under varying conditions.

The implementation of advanced feedback control mechanisms must consider the practical limitations of process measurements and control actions. Sensor placement strategies, measurement frequency, and signal processing techniques all play crucial roles in the effectiveness of feedback control systems. The control system must handle measurement noise and process disturbances effectively while maintaining reliable performance under varying operational conditions. This requires careful consideration of filtering techniques and control algorithm design to ensure robust performance in real-world applications.

Integration with higher-level control and optimization systems represents a critical aspect of advanced feedback control implementation. The feedback control mechanisms must coordinate effectively with supervisory control layers and optimization systems to ensure coherent operation and optimal process performance. This integration enables comprehensive process control while maintaining operational stability and regulatory compliance.

3.6.4 Model Predictive Control Strategies

Model predictive control (MPC) represents one of the most sophisticated approaches to process control in modern PAT systems, enabling predictive optimization of process operations based on dynamic process models. These advanced control strategies utilize detailed process models to predict future system behavior and determine optimal control actions that satisfy multiple objectives and constraints simultaneously. The implementation of MPC strategies requires careful consideration of model accuracy, computational requirements, and control objectives to ensure effective performance in real-world applications (Figure 3.6).

The development of effective MPC strategies relies heavily on the quality and accuracy of underlying process models. These models must capture both steady-state and dynamic behavior accurately while remaining computationally tractable for

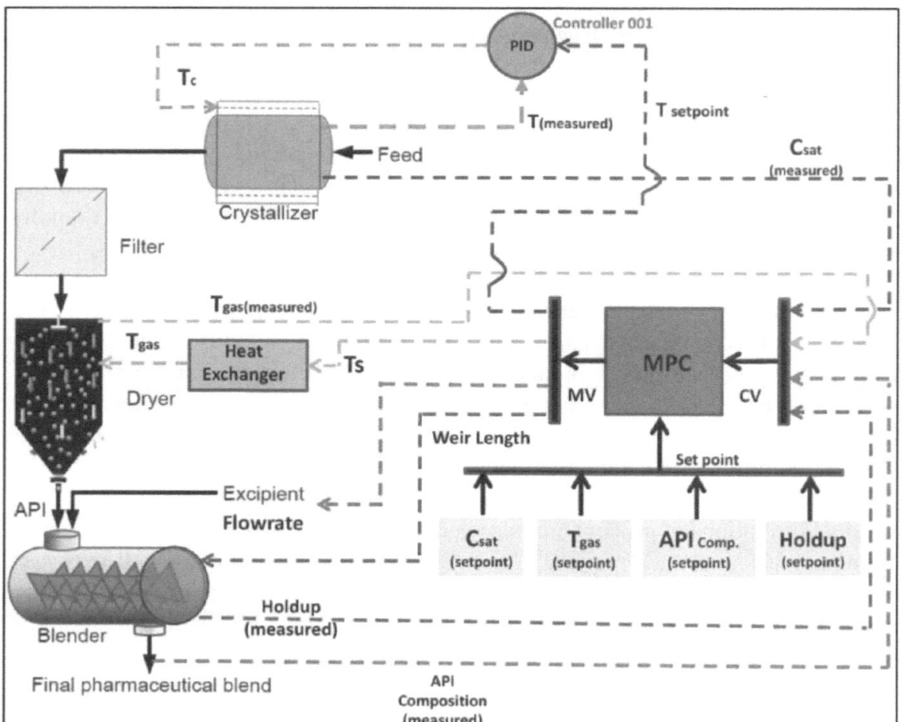

FIGURE 3.6 Integration of model predictive control within PAT framework.

real-time applications. Model development and validation represent critical activities in MPC implementation, requiring sophisticated system identification techniques and extensive validation testing to ensure reliable performance.

The computational aspects of MPC implementation present significant challenges in real-time applications. The optimization algorithms must execute quickly enough to provide timely control actions while maintaining solution quality and reliability. Advanced numerical methods and optimization techniques enable efficient solution of complex control problems while ensuring robust performance under varying operational conditions. The integration of these computational methods with real-time control systems requires careful consideration of hardware capabilities and system architecture.

Constraint handling represents a critical aspect of MPC implementation in PAT systems. The control strategy must handle multiple types of constraints effectively, including equipment limitations, product quality specifications, and safety requirements. Advanced constraint handling techniques enable comprehensive management of operational limits while maintaining optimal process performance. The integration of constraint handlers with optimization algorithms ensures reliable operation while maximizing process efficiency.

3.6.5 Distributed Control System Integration

The integration of distributed control systems (DCSs) within the PAT framework represents a fundamental aspect of modern process control implementation. These sophisticated control architectures distribute control functions across multiple processing units while maintaining coordinated operation through high-speed communication networks. The implementation of distributed control strategies requires careful consideration of system architecture, communication protocols, and coordination mechanisms to ensure reliable and effective process control.

Modern DCS implementations in PAT applications utilize standardized communication protocols and interfaces to ensure seamless integration between different system components. These communication systems must support both high-speed data transfer for real-time control applications and reliable data storage for historical analysis and regulatory compliance. The implementation of effective communication strategies requires careful consideration of network architecture and protocol selection to ensure reliable system performance.

System reliability and redundancy represent critical considerations in distributed control system implementation. Redundant controllers, communication pathways, and power supplies ensure continuous operation even in the event of component failures. The system architecture must support both routine maintenance activities and emergency response scenarios while maintaining process control integrity. This requires careful planning of system components and communication pathways to enable modular replacement and enhancement of control system elements.

The coordination of multiple control units within a distributed system presents significant challenges in PAT applications. Advanced coordination algorithms ensure coherent operation across different process units while maintaining optimal overall system performance. The implementation of effective coordination strategies requires sophisticated software architectures that can handle complex interactions between different control units while maintaining system stability and performance.

The integration of security measures within distributed control systems represents an increasingly important aspect of PAT implementation. Advanced security protocols protect both control systems and process data from unauthorized access or manipulation. The implementation of effective security measures requires careful consideration of access control, data encryption, and system monitoring to ensure reliable and secure operation.

3.7 CONTINUOUS IMPROVEMENT AND KNOWLEDGE MANAGEMENT

3.7.1 Foundations of Continuous Improvement in PAT Systems

The implementation of continuous improvement methodologies within process analytical technology systems represents a sophisticated approach to manufacturing excellence that transcends traditional quality management paradigms. At its core, continuous improvement in PAT systems encompasses a comprehensive framework that integrates real-time process monitoring, advanced analytics, and systematic knowledge management to drive sustained operational excellence. This sophisticated approach leverages the vast amounts of data generated by PAT systems to identify optimization opportunities, implement process improvements, and validate the effectiveness of implemented changes. The systematic nature of continuous improvement in PAT systems ensures that manufacturing processes evolve continuously, adapting to changing requirements while maintaining optimal performance levels and product quality standards.

The integration of continuous improvement principles within PAT frameworks requires a carefully structured approach that balances technological capabilities with organizational requirements. Manufacturing organizations must establish robust mechanisms for data collection, analysis, and interpretation that enable informed decision-making and process optimization. This comprehensive approach encompasses not only the technical aspects of process improvement but also the organizational and cultural elements that support sustained excellence. The systematic collection and analysis of process data, coupled with advanced analytical capabilities, provides the foundation for identifying improvement opportunities and implementing effective solutions that drive operational excellence (Figure 3.7).

The effectiveness of continuous improvement initiatives in PAT systems relies heavily on the establishment of clear performance metrics and evaluation criteria. Organizations must develop comprehensive frameworks for assessing process performance, product quality, and operational efficiency that align with strategic objectives and regulatory requirements. These evaluation frameworks provide the basis for identifying improvement opportunities, prioritizing intervention strategies, and measuring the impact of implemented changes. The systematic approach to performance measurement ensures that improvement initiatives remain focused on critical

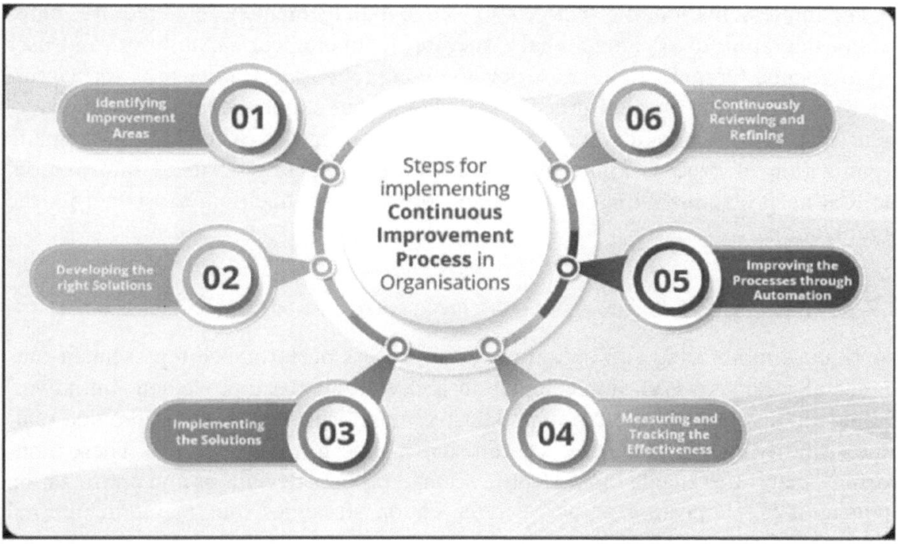

FIGURE 3.7 Integrated continuous improvement cycle in PAT systems.

aspects of process performance while maintaining alignment with organizational goals and objectives.

3.7.2 Knowledge Management Infrastructure

The development of robust knowledge management infrastructure represents a critical element in the successful implementation of PAT systems and continuous improvement initiatives. This infrastructure must support the systematic collection, organization, and dissemination of process knowledge across the organization, enabling informed decision-making and sustained operational excellence. The knowledge management system serves as a central repository for process information, analytical data, and improvement initiatives, providing a comprehensive resource for process optimization and problem-solving activities (Table 3.6).

TABLE 3.6
Components of PAT Knowledge Management Systems

Component	Primary Function	Implementation Level	Key Benefits
Data repository	Information storage	Enterprise	Centralized access
Analysis tools	Data processing	Department	Decision support
Learning systems	Knowledge transfer	Organization	Capability building
Process library	Best practices	Global	Standardization

The implementation of effective knowledge management systems requires careful consideration of organizational structure, technological capabilities, and user requirements. Organizations must develop comprehensive frameworks for capturing, validating, and sharing process knowledge that support continuous improvement initiatives while maintaining data integrity and security. The systematic organization of process knowledge enables rapid access to critical information, facilitating problem-solving activities and process optimization efforts across the organization.

3.7.3 Process Performance Monitoring and Optimization

Continuous monitoring and optimization of process performance represents a fundamental aspect of PAT implementation and continuous improvement initiatives. Organizations must establish comprehensive monitoring systems that provide real-time visibility into critical process parameters and performance metrics. These monitoring capabilities enable rapid identification of process deviations and optimization opportunities, supporting proactive intervention strategies that maintain optimal process performance and product quality.

The implementation of effective performance monitoring systems requires careful consideration of measurement methodologies, data collection frequencies, and analysis techniques. Organizations must develop robust frameworks for data collection and analysis that provide meaningful insights into process performance while maintaining operational efficiency. The systematic monitoring of process parameters enables early detection of potential issues, facilitating rapid intervention and continuous optimization of manufacturing operations.

3.7.4 Quality Systems Integration

The integration of quality management systems within the PAT framework represents a critical element in ensuring sustained operational excellence and regulatory compliance. Organizations must develop comprehensive quality systems that align with PAT implementation strategies while maintaining compliance with relevant regulatory requirements. These integrated quality systems provide the foundation for consistent product quality and process performance while supporting continuous improvement initiatives across the organization.

The development of integrated quality systems requires careful consideration of regulatory requirements, organizational capabilities, and operational objectives. Organizations must establish robust mechanisms for quality control, documentation, and compliance management that support PAT implementation while maintaining regulatory compliance. The systematic integration of quality systems enables consistent product quality while facilitating continuous improvement initiatives across the manufacturing operation.

3.7.5 Training and Capability Development

The development of comprehensive training programs and capability-building initiatives represents a crucial element in the successful implementation of PAT systems and continuous improvement methodologies. Organizations must establish robust training frameworks that enable effective knowledge transfer and skill development across all levels of the organization. These training programs ensure that personnel possess the necessary capabilities to operate PAT systems effectively while contributing to continuous improvement initiatives.

The implementation of effective training programs requires careful consideration of learning objectives, delivery methodologies, and assessment criteria. Organizations must develop comprehensive training materials and assessment tools that enable effective knowledge transfer while validating learning outcomes. The systematic approach to capability development ensures that personnel possess the necessary skills and knowledge to support PAT implementation and continuous improvement initiatives effectively.

3.7.6 Documentation and Change Management

The establishment of robust documentation systems and change management processes represents a critical element in maintaining effective PAT systems and supporting continuous improvement initiatives. Organizations must develop comprehensive frameworks for documenting process knowledge, improvement initiatives, and operational changes that maintain compliance with regulatory requirements while supporting operational excellence. These documentation systems provide the foundation for knowledge retention and transfer while enabling effective change management across the organization.

The implementation of effective documentation systems requires careful consideration of regulatory requirements, operational needs, and user accessibility. Organizations must establish robust mechanisms for document control, version management, and change tracking that maintain compliance while supporting operational efficiency. The systematic approach to documentation and change management ensures that process knowledge is effectively captured and maintained while enabling controlled implementation of improvement initiatives.

3.8 ARTIFICIAL INTELLIGENCE IN PROCESS ANALYTICAL TECHNOLOGY

3.8.1 Integration of AI Systems in PAT Framework

The integration of artificial intelligence within process analytical technology represents a revolutionary advancement in manufacturing process control and optimization. Modern industrial processes generate unprecedented volumes of data from multiple sensors, analyzers, and control systems, creating a complex ecosystem of information that traditional analysis methods struggle to process effectively.

Artificial intelligence systems, particularly machine learning algorithms, have emerged as powerful tools for extracting meaningful insights from this vast data landscape. These advanced computational methods enable real-time process optimization, predictive maintenance, and quality control at a level of sophistication previously unattainable through conventional methods.

The implementation of AI systems within the PAT framework involves multiple layers of technological integration, ranging from data preprocessing and feature extraction to advanced model development and deployment. Modern manufacturing facilities utilize sophisticated sensor networks that continuously monitor various process parameters, generating massive datasets that contain valuable information about process behavior and product quality. Artificial intelligence systems can process this high-dimensional data in real time, identifying complex patterns and relationships that might be invisible to human operators or traditional statistical methods. This capability enables more nuanced process control and optimization strategies, leading to improved product quality and operational efficiency.

The architectural framework for AI integration in PAT systems encompasses several key components that work in harmony to enable intelligent process control. At the foundation lies the data acquisition layer, where multiple sensors and analyzers collect real-time process information. This raw data undergoes preliminary processing through edge computing devices, which perform initial filtering and feature extraction. The processed data then flows into advanced AI systems that employ various machine learning algorithms to analyze process patterns, predict potential issues, and generate optimal control strategies. This hierarchical approach ensures efficient data processing while maintaining the real-time responsiveness required for effective process control.

The implementation of AI systems in PAT requires careful consideration of various technical and operational factors. System architects must balance the computational requirements of advanced AI algorithms with the need for real-time process control. This balance often necessitates the use of distributed computing architectures, where different aspects of data processing and analysis are distributed across multiple computational nodes. Edge computing devices handle immediate process control requirements, while more complex analytical tasks are performed on centralized servers or cloud platforms. This distributed approach ensures optimal system performance while maintaining the reliability and responsiveness required for critical process control applications.

3.8.2 Machine Learning Applications in Process Control

Machine learning algorithms have revolutionized process control strategies within the PAT framework by enabling more sophisticated approaches to process optimization and quality control. These advanced computational methods can automatically identify complex relationships between process parameters and product quality attributes, enabling more precise control strategies that adapt to changing process conditions. Deep learning networks, in particular, have demonstrated remarkable capabilities in handling the complex, nonlinear relationships common in manufacturing processes, enabling more accurate predictions and control decisions than traditional statistical methods.

The application of machine learning in process control extends beyond simple parameter optimization to encompass comprehensive process understanding and optimization. Neural networks and other advanced algorithms can analyze historical process data to identify optimal operating conditions, predict potential quality issues before they occur, and suggest corrective actions to maintain product quality. This predictive capability enables proactive process control strategies that can prevent quality deviations before they impact product specifications, resulting in improved process efficiency and reduced waste.

Reinforcement learning algorithms have emerged as particularly powerful tools for process optimization in PAT systems. These algorithms can learn optimal control strategies through continuous interaction with the process, adapting their behavior based on feedback from process outcomes. This adaptive learning capability enables the development of control systems that can optimize process performance under varying conditions, leading to improved efficiency and product quality. The implementation of reinforcement learning in PAT systems represents a significant advancement in process control technology, enabling more sophisticated approaches to process optimization than previously possible (Table 3.7).

3.8.3 Advanced Analytics and Decision Support Systems

The integration of advanced analytics within PAT systems has created new paradigms for process understanding and control. Modern analytical platforms combine traditional statistical methods with machine learning algorithms to provide comprehensive insights into process behavior and product quality. These advanced analytical capabilities enable process engineers to identify subtle patterns and relationships in process data that might indicate potential quality issues or optimization opportunities. The combination of real-time analysis and historical data mining creates a powerful platform for process optimization and quality control.

The implementation of decision support systems within the PAT framework represents a crucial advancement in process management and control. These sophisticated

TABLE 3.7
Machine Learning Applications in PAT Systems

Application Area	ML Algorithm Type	Primary Benefits	Implementation Complexity
Process optimization	Deep learning	Improved quality control	High
Predictive maintenance	Random forests	Reduced downtime	Medium
Quality prediction	Neural networks	Enhanced product consistency	High
Fault detection	SVM/clustering	Early warning systems	Medium

systems combine real-time process data with historical information and expert knowledge to provide operators and engineers with actionable insights for process optimization. The integration of machine learning algorithms enhances the capability of these systems to identify complex patterns and relationships in process data, enabling more informed decision-making processes. Advanced analytics platforms can process multiple data streams simultaneously, considering both direct process measurements and derived parameters to provide a comprehensive understanding of process behavior and product quality dynamics.

Modern decision support systems in PAT applications utilize sophisticated visualization techniques to present complex process information in an intuitive and actionable format. These visualization capabilities enable operators to quickly identify process trends, potential issues, and optimization opportunities without requiring extensive data analysis expertise. The combination of advanced analytics and intuitive visualization creates a powerful platform for process optimization and quality control, enabling operators to make informed decisions based on comprehensive process understanding rather than limited snapshot views of process parameters.

The integration of historical process data with real-time measurements enables more sophisticated approaches to process optimization and control. Advanced analytics systems can identify correlations between historical process conditions and product quality outcomes, enabling the development of predictive models that anticipate potential quality issues before they occur. This predictive capability enables proactive process control strategies that can prevent quality deviations before they impact product specifications, resulting in improved process efficiency and reduced waste generation.

Knowledge management systems play a crucial role in enhancing the effectiveness of decision support platforms in PAT applications. These systems capture and organize process knowledge, operational experience, and best practices, making this information readily available to operators and engineers when needed. The integration of machine learning algorithms enables these knowledge management systems to continuously learn from operational experience, improving their ability to provide relevant recommendations and insights over time.

3.8.4 Real-Time Process Optimization Through AI

The implementation of real-time process optimization systems represents one of the most significant applications of artificial intelligence in PAT environments. These advanced systems continuously monitor process conditions and product quality parameters using sophisticated algorithms to identify optimal operating conditions that maximize product quality while minimizing resource consumption. The ability to perform complex optimization calculations in real time enables more dynamic approaches to process control, allowing manufacturing systems to adapt quickly to changing conditions while maintaining optimal performance.

The development of effective real-time optimization strategies requires careful consideration of multiple factors including process dynamics, constraint handling, and computational efficiency. Advanced optimization algorithms must balance the need for optimal process performance with practical constraints such as equipment limitations, safety requirements, and product quality specifications. Machine learning algorithms play a crucial role in this optimization process by enabling more sophisticated approaches to constraint handling and process modeling than possible with traditional optimization methods.

Implementation of real-time optimization systems requires sophisticated software architectures that can handle the computational demands of complex optimization algorithms while maintaining the responsiveness required for effective process control. Modern systems often employ distributed computing architectures where optimization calculations are performed across multiple computational nodes, enabling more efficient processing of complex optimization problems while maintaining real-time performance requirements.

3.8.5 Predictive Maintenance and Fault Detection

The application of artificial intelligence in predictive maintenance represents a crucial advancement in PAT system capabilities. Advanced machine learning algorithms can analyze complex patterns in equipment performance data to identify potential maintenance issues before they lead to equipment failure or quality problems. This predictive capability enables more proactive approaches to maintenance planning, reducing unplanned downtime while optimizing maintenance resource allocation.

Fault detection systems enhanced by artificial intelligence can identify subtle changes in process behavior that might indicate developing problems or equipment degradation. These systems analyze multiple data streams simultaneously, considering both direct measurements and derived parameters to provide early warning of potential issues. The ability to detect problems early enables maintenance teams to address issues before they impact product quality or cause equipment damage, resulting in improved operational reliability and reduced maintenance costs.

The integration of predictive maintenance systems with overall equipment effectiveness (OEE) monitoring creates comprehensive platforms for equipment reliability management. These integrated systems provide maintenance teams with detailed insights into equipment performance trends, enabling more effective planning of maintenance activities and resource allocation. The combination of predictive capabilities with detailed performance analysis enables more sophisticated approaches to maintenance strategy development and implementation.

3.8.6 Quality Control and Process Validation

Artificial intelligence systems have revolutionized quality control processes within the PAT framework by enabling more sophisticated approaches to product quality assessment and process validation. Advanced machine learning algorithms can analyze complex relationships between process parameters and product quality attributes, enabling more accurate prediction of product quality outcomes. This predictive capability enables more proactive approaches to quality control, allowing manufacturers to identify and address potential quality issues before they impact final product specifications.

The implementation of AI-enhanced quality control systems requires sophisticated approaches to data validation and model verification. These systems must maintain high levels of accuracy and reliability while processing complex data streams in real time. The development of effective validation strategies ensures that quality predictions remain accurate and reliable over time, even as process conditions and product requirements evolve.

The integration of quality control systems with process validation protocols creates comprehensive platforms for quality assurance and regulatory compliance. These integrated systems provide detailed documentation of process performance and product quality parameters, enabling more efficient approaches to regulatory compliance and quality system management. The combination of advanced analytics with comprehensive documentation capabilities ensures that manufacturing processes maintain consistent compliance with regulatory requirements while optimizing product quality and process efficiency.

3.8.7 Future Trends and Emerging Technologies

The evolution of artificial intelligence applications in PAT systems continues to be driven by rapid advances in computing technology and algorithm development. Emerging technologies such as quantum computing and advanced neural network architectures promise to enable even more sophisticated approaches to process optimization and control. These technological advances will enable manufacturing systems to handle increasingly complex optimization problems while maintaining the real-time performance required for effective process control.

The integration of edge computing technologies with artificial intelligence systems represents an important trend in PAT system development. Edge computing enables more efficient processing of sensor data by performing initial analysis close to the data source, reducing the load on central computing systems while improving system responsiveness. This distributed approach to data processing enables more sophisticated approaches to process control while maintaining the reliability and performance required for critical manufacturing applications.

The development of more sophisticated approaches to model interpretability and validation represents another important trend in AI applications for PAT systems. As machine learning models become more complex, ensuring their predictions remain interpretable and reliable becomes increasingly important. Advanced techniques for model validation and interpretation enable more effective use of AI systems in regulated manufacturing environments where decisions must be fully traceable and justifiable.

3.9 ENHANCING DATA ACQUISITION AND PROCESS CONTROL THROUGH MACHINE LEARNING ALGORITHMS

3.9.1 Advanced Data Acquisition Frameworks

The implementation of machine learning algorithms in process analytical echnology has fundamentally transformed traditional approaches to data acquisition and processing. Modern manufacturing environments generate massive volumes of process data from diverse sources, creating complex challenges for traditional data handling methods. Advanced machine learning frameworks enable more sophisticated approaches to data acquisition, incorporating intelligent filtering, feature extraction, and data validation capabilities that enhance the quality and utility of process information. These systems can automatically identify relevant process features and patterns while filtering out noise and irrelevant information, significantly improving the efficiency of data collection and analysis processes.

The development of intelligent sensor networks represents a crucial advancement in PAT data acquisition capabilities. These sophisticated networks incorporate machine learning algorithms at multiple levels, from individual sensor nodes to centralized data processing systems. Edge computing devices equipped with machine learning capabilities can perform preliminary data analysis and feature extraction directly at the sensor level, reducing data transmission requirements while maintaining information quality. This distributed approach to data processing enables more efficient utilization of communication bandwidth while ensuring the capture of critical process information.

Real-time data validation and reconciliation represent critical aspects of modern PAT implementations. Machine learning algorithms can automatically identify measurement anomalies and sensor failures, enabling more reliable data collection and analysis. Advanced validation techniques incorporate historical process knowledge and physical constraints to ensure measurement consistency and reliability. The integration of these validation capabilities with data acquisition systems creates robust platforms for process monitoring and control.

3.9.2 Intelligent Process Control Integration

The integration of machine learning algorithms with process control systems represents a significant advancement in PAT implementation capabilities. These advanced systems can automatically identify complex relationships between process parameters and product quality attributes, enabling more sophisticated approaches to process control and optimization. The ability to learn from historical process data and adapt control strategies based on operating experience creates more robust and effective control systems than possible with traditional methods.

Contemporary process control systems enhanced by machine learning capabilities demonstrate remarkable adaptability to changing process conditions. These systems can automatically adjust control parameters and strategies based on observed process behavior, maintaining optimal performance even as operating conditions evolve. The integration of learning capabilities with traditional control algorithms creates hybrid systems that combine the reliability of established control methods with the adaptability of machine learning approaches (Figure 3.8).

The implementation of intelligent control systems requires sophisticated software architectures that can handle both traditional control functions and machine learning capabilities. These architectures must support real-time processing of process data while maintaining the reliability and responsiveness required for critical control

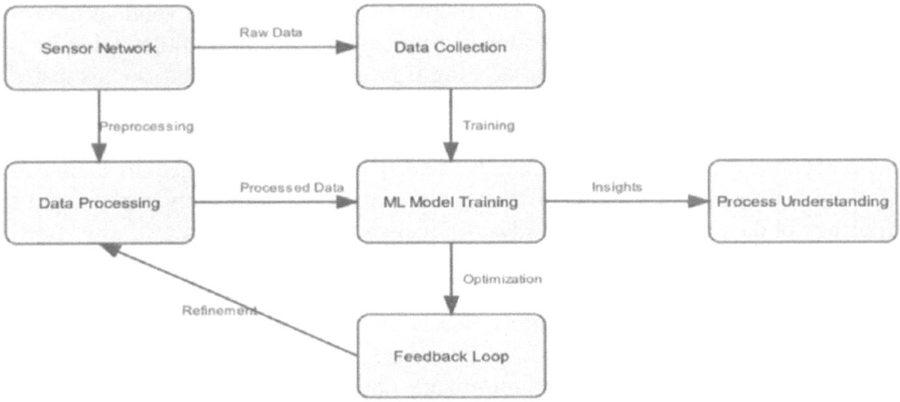

FIGURE 3.8 Machine learning-enhanced data acquisition framework: From sensor networks to process understanding.

applications. The development of effective system architectures requires careful consideration of computational requirements, communication protocols, and integration strategies.

3.9.3 Advanced Pattern Recognition and Feature Extraction

Pattern recognition capabilities enhanced by machine learning algorithms enable more sophisticated approaches to process monitoring and control. These systems can automatically identify complex patterns in process data that might indicate developing problems or optimization opportunities. The ability to recognize subtle patterns and relationships in process data enables more proactive approaches to process control and quality assurance.

Feature extraction techniques based on machine learning algorithms enable more efficient processing of complex process data. These methods can automatically identify relevant process characteristics while filtering out noise and irrelevant information. The development of effective feature extraction strategies requires careful consideration of process dynamics and measurement characteristics to ensure reliable performance.

The integration of advanced pattern recognition capabilities with process control systems enables more sophisticated approaches to fault detection and process optimization. These integrated systems can automatically identify potential problems before they impact product quality, enabling more proactive approaches to process control and maintenance planning.

3.9.4 Adaptive Learning Systems

The implementation of adaptive learning systems represents a crucial advancement in PAT capabilities. These sophisticated systems can continuously update their knowledge and control strategies based on operational experience, enabling more effective process optimization and control. The ability to learn from process data and adapt to changing conditions creates more robust and efficient manufacturing systems.

Modern adaptive learning systems incorporate multiple learning mechanisms to handle different aspects of process control and optimization. Supervised learning algorithms enable the development of accurate predictive models based on historical process data, while reinforcement learning approaches enable the optimization of control strategies through continuous interaction with the process. The integration of multiple learning approaches creates comprehensive systems that can handle both routine operations and special situations effectively.

The development of effective adaptive learning systems requires careful consideration of learning algorithm selection and implementation strategies. These systems must maintain reliable performance while continuously updating their knowledge and control strategies. The implementation of appropriate validation and verification procedures ensures that learning systems maintain accurate and reliable operation over time.

3.10 CONCLUSION

Process analytical technology represents a transformative approach to manufacturing process control and optimization, incorporating advanced analytical methods, sophisticated control strategies, and emerging technologies to achieve unprecedented levels of process understanding and control. The integration of artificial intelligence and machine learning capabilities has further enhanced these systems, enabling more sophisticated approaches to process optimization and quality assurance. The comprehensive framework provided by PAT creates powerful platforms for manufacturing excellence, enabling continuous improvement and innovation in process control and optimization.

The evolution of PAT systems continues to be driven by advances in technology and analytical capabilities. The integration of artificial intelligence and machine learning algorithms has created new opportunities for process optimization and control, enabling more sophisticated approaches to manufacturing excellence. The development of more advanced sensor technologies, analytical methods, and control strategies will continue to enhance PAT capabilities, enabling even greater levels of process understanding and control in the future.

The implementation of PAT systems requires careful consideration of multiple factors, including technology selection, system integration, and regulatory compliance. The success of PAT implementations depends heavily on the effective integration of various system components and the development of appropriate operational strategies. Organizations must carefully consider their specific requirements and constraints when implementing PAT systems to ensure optimal performance and value realization.

Looking forward, the continued evolution of PAT systems will likely incorporate emerging technologies such as quantum computing, advanced materials science, and novel analytical methods. These technological advances will enable even more sophisticated approaches to process understanding and control, creating new opportunities for manufacturing excellence and innovation. The integration of these emerging technologies with established PAT frameworks will drive continued advancement in manufacturing process control and optimization.

The future of manufacturing excellence lies in the continued development and implementation of advanced PAT systems. These sophisticated platforms enable unprecedented levels of process understanding and control, creating new opportunities for innovation and operational excellence. As technology continues to evolve, PAT systems will incorporate new capabilities and methodologies, enabling even greater levels of manufacturing performance and product quality assurance.

4 Standard Operating Procedures in Biotechnology

4.1 INTRODUCTION

The landscape of biotechnology manufacturing has undergone remarkable transformation since its inception, evolving from small-scale laboratory procedures to highly sophisticated industrial processes capable of producing therapeutic proteins, vaccines, and other biological products at commercial scale. At the heart of this evolution lies the critical importance of standard operating procedures (SOPs), which have transcended their original role as simple operational guidelines to become comprehensive systems that integrate process control, quality assurance, and regulatory compliance. The modern biotechnology facility represents a complex ecosystem where multiple SOPs must function in perfect harmony to ensure product quality, process efficiency, and patient safety. These procedures embody not only the technical requirements for successful manufacturing but also the accumulated knowledge and experience of countless scientists, engineers, and regulatory experts who have contributed to the field's development.

4.1.1 Regulatory Framework and Quality Systems Integration

The regulatory landscape surrounding biotechnology manufacturing has become increasingly complex, with regulatory bodies worldwide implementing stringent requirements for process validation, product quality, and documentation. SOPs serve as the primary mechanism for demonstrating compliance with Good Manufacturing Practice (GMP) guidelines, providing detailed documentation of every critical process step and associated control parameters. This comprehensive documentation system extends beyond basic operational instructions to encompass risk assessment protocols, deviation management procedures, and continuous improvement initiatives. The integration of quality systems within SOPs has become particularly crucial as the industry moves toward real-time release testing and continuous manufacturing paradigms. Modern SOPs must therefore balance the need for rigid control with sufficient flexibility to accommodate process optimization and technological advancement (Figure 4.1).

DOI: 10.1201/9781003624592-4

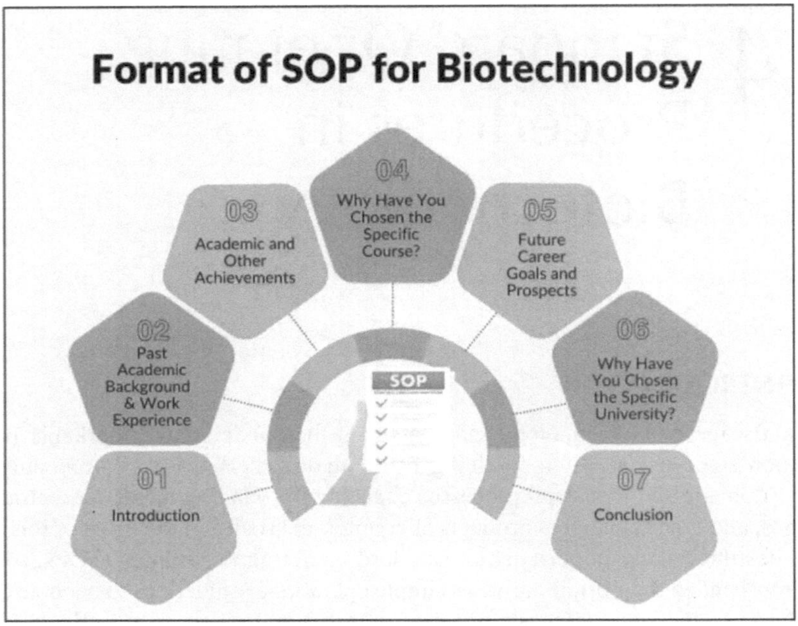

FIGURE 4.1 Integrated framework of modern biotechnology SOPs.

4.1.2 Technological Advancement and Process Control Strategies

The unprecedented advancement in biotechnology processing equipment and analytical technologies has revolutionized the implementation of SOPs in manufacturing environments. The adoption of single-use technologies, automated control systems, and advanced analytical methods has necessitated the development of more sophisticated procedures that can accommodate these technological innovations while maintaining process consistency and product quality. Process analytical technology (PAT) initiatives have become increasingly integrated within SOP frameworks, enabling real-time monitoring and control of critical process parameters. This evolution toward smart manufacturing systems has been particularly evident in the implementation of continuous processing platforms, where traditional batch-based procedures must be adapted to support continuous operation and verification.

4.1.3 Economic Implications and Global Manufacturing Networks

The economic impact of well-designed and effectively implemented SOPs extends far beyond basic operational efficiency. In an industry where a single batch of biological product can represent millions of dollars in value, the importance of robust procedures becomes paramount. SOPs serve as critical risk mitigation tools, helping to prevent costly manufacturing deviations while ensuring consistent product quality across multiple production sites. The globalization of biotechnology manufacturing has further emphasized the importance of comprehensive SOPs in facilitating technology transfer between different facilities while maintaining consistent quality standards across diverse geographical locations. This global perspective has led to the development of more robust and adaptable procedures that can accommodate regional variations in regulatory requirements while maintaining the fundamental principles of Good Manufacturing Practice.

4.1.4 Risk Management and Quality by Design Integration

The integration of quality by design (QbD) principles within SOP frameworks represents a significant advancement in biotechnology manufacturing. This systematic approach to process development and control has fundamentally changed how SOPs are structured and implemented, emphasizing the importance of understanding critical quality attributes and their relationship to process parameters. Risk management strategies have become increasingly sophisticated, incorporating elements of failure mode and effects analysis (FMEA), process capability assessments, and continuous monitoring protocols. These advanced approaches to risk management have enabled manufacturers to develop more robust processes while simultaneously reducing the likelihood of batch failures and product quality issues.

4.1.5 Training and Personnel Qualification

The human element remains crucial in the successful implementation of SOPs, despite increasing automation and technological advancement. Comprehensive training programs must be developed and maintained to ensure that personnel at all levels understand not only the technical aspects of their assigned procedures but also the underlying principles and potential impacts of their actions on product quality. Modern SOPs must therefore include detailed training requirements, competency assessment protocols, and ongoing evaluation procedures to maintain the highest standards of operational excellence. This emphasis on personnel qualification extends beyond basic operational training to encompass understanding of quality systems, regulatory requirements, and risk management principles.

4.1.6 Documentation Systems and Knowledge Management

The management of SOPs in modern biotechnology facilities represents a significant challenge, requiring sophisticated document control systems and knowledge management platforms. Electronic documentation systems have become increasingly prevalent, enabling more efficient review and approval processes while maintaining the integrity of controlled documents. These systems must accommodate the complex relationships between different procedures while ensuring that all relevant personnel have access to the most current versions of applicable documents. The integration of electronic batch records, automated data collection systems, and laboratory information management systems (LIMS) has further complicated the documentation landscape, requiring careful consideration of data integrity and system validation requirements (Table 4.1).

TABLE 4.1
Critical Elements of Modern Biotechnology SOPs

Element	Primary Function	Impact on Manufacturing	Regulatory Significance
Process documentation	Standardization of procedures	Ensures consistency	Demonstrates GMP compliance
Quality control	Product quality verification	Maintains specifications	Supports batch release
Risk management	Prevention of deviations	Reduces failures	Ensures patient safety
Training requirements	Personnel qualification	Maintains excellence	Documents competency
Change control	Process modification management	Preserves validity	Maintains compliance
Technology integration	Automation and control	Enhances efficiency	Supports data integrity
Knowledge management	Information organization	Facilitates improvements	Enables regulatory reporting
Continuous improvement	Process optimization	Increases productivity	Demonstrates commitment to quality

The subsequent sections of this chapter will explore in detail the specific applications of SOPs across different aspects of biotechnology manufacturing, from upstream processing through final product preparation. Through examination of case studies and practical examples, we will demonstrate how these fundamental principles are applied to ensure the consistent production of high-quality biological products while maintaining compliance with regulatory requirements and industry best practices.

4.2 UPSTREAM PROCESSING OF PROTEINS

4.2.1 Fundamentals of Cell Line Development and Selection

The foundation of successful protein production lies in the meticulous development and selection of appropriate cell lines, a process that demands extensive expertise in cellular biology, genetic engineering, and process optimization. Cell line development represents the critical initial phase of upstream processing, where the production efficiency and product quality characteristics are largely predetermined. Modern biotechnology facilities employ sophisticated screening methodologies that combine high-throughput analysis with advanced cell culture techniques to identify and isolate high-performing cell lines. The selection process must consider multiple factors simultaneously, including specific productivity, growth characteristics, genetic stability, and the cell line's ability to produce proteins with desired post-translational modifications.

The journey of cell line development begins with the careful selection of host cell types, with Chinese Hamster Ovary (CHO) cells remaining the predominant choice for therapeutic protein production due to their exceptional ability to produce proteins with human-compatible glycosylation patterns. The host cell selection process extends beyond basic growth characteristics to encompass considerations of regulatory acceptance, historical safety profiles, and the availability of established platforms for genetic modification. Contemporary approaches to cell line development increasingly incorporate automated systems for cell screening and characterization, enabling the evaluation of thousands of potential candidates simultaneously while maintaining detailed documentation of selection criteria and clone histories (Figure 4.2).

The implementation of stable gene expression systems represents another critical aspect of cell line development, where multiple strategies must be carefully evaluated to achieve optimal protein production. Modern approaches utilize advanced vector design principles that incorporate elements such as matrix attachment regions (MARs), scaffold/matrix attachment regions (S/MARs), and other regulatory elements to enhance expression stability and reduce the likelihood of gene silencing over extended cultivation periods. The integration of these elements must be carefully controlled and documented within standard operating procedures to ensure reproducibility and regulatory compliance.

Selection and amplification systems play a crucial role in establishing high-producing cell lines, with methodologies evolving beyond traditional approaches to incorporate novel selection markers and amplification strategies. The development of selection protocols must balance the desire for high productivity with considerations of cell line stability and process robustness. Modern cell line development platforms increasingly utilize multiple selection markers and amplification systems in combination, requiring careful optimization of selection pressure and amplification conditions to achieve desired expression levels while maintaining cell line stability (Table 4.2).

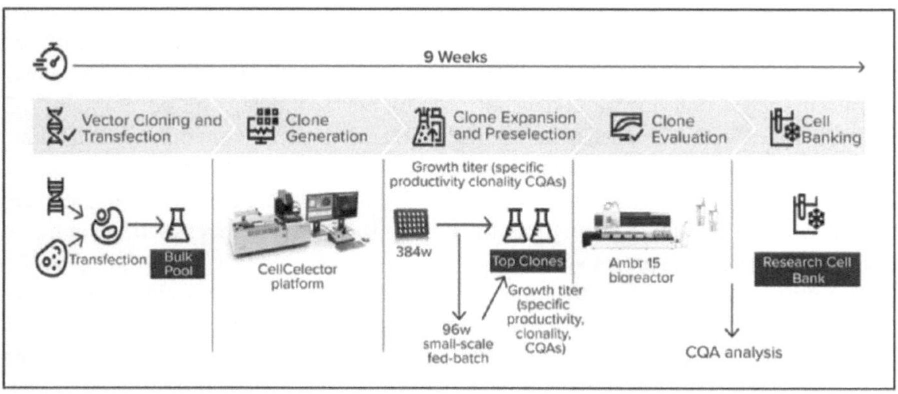

FIGURE 4.2 Integrated cell line development platform.

TABLE 4.2
Critical Parameters in Cell Line Development

Parameter	Impact on Process	Control Strategy	Quality Implications
Expression stability	Long-term productivity	Genetic elements selection	Product consistency
Growth characteristics	Process efficiency	Media optimization	Resource utilization
Product quality	Therapeutic efficacy	Clone selection	Regulatory compliance
Genetic stability	Manufacturing reliability	Characterization methods	Process validation
Scale-up potential	Commercial viability	Platform evaluation	Technology transfer

4.2.2 Media Development and Optimization Strategies

The development of optimal cell culture media represents a cornerstone of successful upstream processing, requiring deep understanding of cellular metabolism, nutrient requirements, and process optimization principles. Modern media development strategies have evolved significantly from traditional approaches, incorporating chemically defined components and sophisticated supplementation strategies to achieve optimal cell growth and protein production. The complexity of media development extends beyond basic nutritional requirements to encompass considerations of product quality attributes, process consistency, and regulatory compliance.

The transition toward chemically defined media has revolutionized upstream processing, enabling greater process control and reproducibility while simultaneously reducing regulatory concerns associated with animal-derived components. Contemporary media development programs utilize systematic approaches to optimize basal media composition, feed strategies, and supplementation protocols. This optimization process must consider not only cell growth and protein production but also the impact of media components on product quality attributes such as glycosylation patterns and protein aggregation tendencies.

Media optimization strategies increasingly incorporate principles of quality by design (QbD), utilizing design of experiments (DoE) approaches to establish robust operating spaces that ensure consistent process performance. The development of feeding strategies represents a particularly critical aspect of media optimization, requiring careful balance of nutrient provision, metabolic waste accumulation, and process economics. Modern approaches often utilize dynamic feeding strategies that adapt to cellular demands throughout the cultivation process, necessitating sophisticated monitoring and control systems.

The impact of media composition on product quality attributes cannot be overstated, with seemingly minor variations in media components potentially leading to significant changes in protein characteristics. Contemporary media development programs must therefore incorporate extensive analysis of critical quality attributes throughout the optimization process, ensuring that improvements in cell growth and productivity do not come at the expense of product quality. This analysis extends beyond basic protein quantification to include detailed characterization of post-translational modifications, aggregation profiles, and biological activity.

Process scalability considerations must be integrated into media development programs from the earliest stages, ensuring that optimization strategies developed at small scale can be effectively implemented in commercial manufacturing processes. This scalability assessment must consider not only the technical aspects of media preparation and delivery but also economic considerations related to raw material costs and supply chain reliability. The development of robust media preparation procedures represents a critical aspect of process validation, requiring careful attention to mixing conditions, component stability, and sterility assurance.

4.2.3 Bioreactor Design and Operation Principles

The successful implementation of upstream protein production processes relies heavily on sophisticated bioreactor systems that provide optimal conditions for cellular growth and protein expression. Modern bioreactor design incorporates advanced engineering principles to ensure homogeneous conditions throughout the culture volume while maintaining precise control over critical process parameters. The evolution of bioreactor technology has led to the development of various configurations, from traditional stirred-tank designs to novel single-use systems, each offering specific advantages for different applications within the biotechnology industry. The selection of appropriate bioreactor systems must consider not only immediate process requirements but also long-term manufacturing strategies, scalability considerations, and regulatory compliance requirements.

Contemporary bioreactor designs increasingly incorporate sophisticated sensor systems that enable real-time monitoring of multiple process parameters simultaneously. These monitoring systems extend beyond traditional measurements of pH, dissolved oxygen, and temperature to include advanced sensors for metabolite analysis, cell density determination, and product quality assessment. The integration of these monitoring systems with automated control platforms has revolutionized bioreactor operation, enabling precise maintenance of optimal conditions throughout the cultivation process. This level of control becomes particularly critical during extended cultivation periods, where maintaining consistent conditions directly impacts product quality and process productivity.

The implementation of effective mixing strategies represents a fundamental aspect of bioreactor design and operation, requiring careful consideration of fluid dynamics, mass transfer requirements, and shear sensitivity of cultured cells. Modern bioreactor systems utilize sophisticated impeller designs and mixing configurations that achieve optimal homogeneity while minimizing shear stress on cultured cells. The development of mixing strategies must balance the need for effective mass transfer with considerations of energy input, foam formation, and scale-up implications. This balance becomes particularly critical in large-scale manufacturing operations, where maintaining consistent conditions throughout large culture volumes presents significant engineering challenges (Figure 4.3).

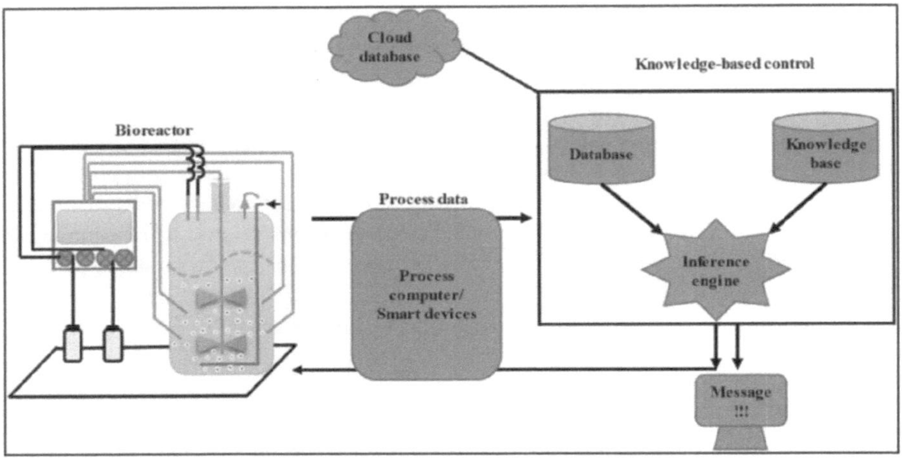

FIGURE 4.3 Advanced bioreactor control systems.

The evolution of single-use bioreactor systems has introduced new considerations in process design and operation, requiring careful evaluation of materials compatibility, extractables and leachables profiles, and system reliability. While offering significant advantages in terms of operational flexibility and contamination risk reduction, single-use systems present unique challenges in terms of process control and scalability. The implementation of single-use technology must consider not only immediate process requirements but also long-term manufacturing strategies and sustainability objectives. The development of robust change-over procedures and system integrity testing protocols becomes particularly critical when implementing single-use technologies in GMP manufacturing environments.

Gas transfer optimization represents another critical aspect of bioreactor design and operation, requiring sophisticated approaches to ensure adequate oxygen supply while managing carbon dioxide accumulation. Modern bioreactor systems incorporate advanced sparging strategies and gas flow control systems that adapt to changing cellular demands throughout the cultivation process. The development of effective gas transfer strategies must consider not only immediate cellular requirements but also the impact of gas flow patterns on foam formation, cell shear sensitivity, and process scalability. This optimization becomes particularly critical in high-cell-density processes, where oxygen demand and carbon dioxide accumulation can significantly impact process performance.

4.2.4 Process Monitoring and Control Strategies

The implementation of effective process monitoring and control strategies represents a cornerstone of successful upstream protein production, requiring sophisticated approaches to data collection, analysis, and process adjustment. Modern biotechnology facilities utilize advanced process analytical technology (PAT) initiatives that enable comprehensive monitoring of critical process parameters and quality attributes throughout the cultivation process. The development of monitoring strategies must consider not only immediate process control requirements but also long-term trends analysis and process optimization objectives. The integration of multiple monitoring systems creates complex data streams that require sophisticated analysis approaches to extract meaningful process information and guide control decisions.

Real-time monitoring capabilities have evolved significantly, incorporating advanced sensor technologies that enable continuous assessment of critical process parameters without compromising sterility or process integrity. These monitoring systems extend beyond traditional physical parameters to include sophisticated approaches for biochemical analysis, cell concentration determination, and product quality assessment. The implementation of advanced monitoring strategies requires careful consideration of sensor placement, calibration protocols, and maintenance requirements to ensure reliable data collection throughout extended cultivation periods. The development of robust calibration and verification procedures becomes particularly critical when implementing advanced monitoring systems in GMP manufacturing environments.

The evolution of control strategies has paralleled advancements in monitoring capabilities, with modern systems utilizing sophisticated algorithms that enable predictive control approaches based on real-time process data. These control systems must balance the need for precise parameter maintenance with considerations of process robustness and operational flexibility. The development of control strategies increasingly incorporates artificial intelligence (AI) and machine learning approaches that enable adaptive response to changing process conditions while maintaining consistent product quality. The implementation of advanced control systems requires careful validation to ensure reliable operation across different process conditions and scale ranges (Table 4.3).

Data management systems play an increasingly critical role in process monitoring and control, requiring sophisticated approaches to data collection, storage, and analysis. Modern facilities utilize integrated data management platforms that enable real-time visualization of process parameters while maintaining complete data integrity and traceability. The development of data management strategies must consider not only immediate process control requirements but also long-term trending analysis

TABLE 4.3
Critical Process Parameters and Control Strategies

Parameter Category	Monitoring Approach	Control Strategy	Quality Impact
Physical parameters	Online sensors	Automated feedback	Process stability
Chemical parameters	At-line analysis	Feed rate adjustment	Metabolic balance
Cellular parameters	Real-time imaging	Growth phase control	Product quality
Product quality	Spectroscopic methods	Process adjustment	Final specifications
Metabolic indicators	Online analyzers	Nutrient addition	Cell viability

and regulatory compliance objectives. The implementation of electronic data capture systems requires careful validation to ensure reliable operation and compliance with data integrity requirements.

The integration of process monitoring and control systems with manufacturing execution systems (MESs) represents another critical aspect of modern biotechnology operations. These integrated systems enable comprehensive process documentation while facilitating rapid response to process deviations or quality concerns. The development of integration strategies must consider not only technical compatibility requirements but also operational efficiency and compliance objectives. The implementation of integrated systems requires careful attention to user interface design, access control protocols, and system maintenance requirements to ensure reliable operation in GMP manufacturing environments.

4.3 DOWNSTREAM PROCESSING OF PROTEINS

4.3.1 Introduction to Downstream Processing

Downstream processing represents one of the most critical and complex aspects of biotechnology manufacturing, encompassing all steps required to purify and isolate the desired protein product from the cellular material and culture medium. This sophisticated series of unit operations must be carefully designed and executed to ensure maximum product recovery while maintaining the protein's biological activity and meeting stringent quality requirements. The complexity of downstream processing lies not only in the technical challenges of separation and purification but also in the need to maintain consistent performance across multiple batches while adhering to regulatory guidelines and cost constraints. Modern downstream processing has evolved significantly with the advent of new technologies and materials, enabling higher purification efficiencies and better process control.

The economic significance of downstream processing cannot be overstated, as it typically accounts for 50%–80% of the total production costs in biotechnology manufacturing. This substantial cost contribution stems from the need for specialized equipment, high-quality chromatography resins, filtration materials, and the extensive validation requirements associated with each purification step. Furthermore, the increasing demands for higher product purity, especially in therapeutic protein production, have led to more complex purification trains that must remove an ever-growing list of process and product-related impurities. The development of robust downstream processes therefore requires careful consideration of both technical and economic factors to ensure commercial viability while maintaining product quality (Figure 4.4).

4.3.2 Primary Recovery Operations

4.3.2.1 Cell Separation and Clarification

The initial phase of downstream processing begins with the separation of cells and cellular debris from the culture medium containing the product of interest. This

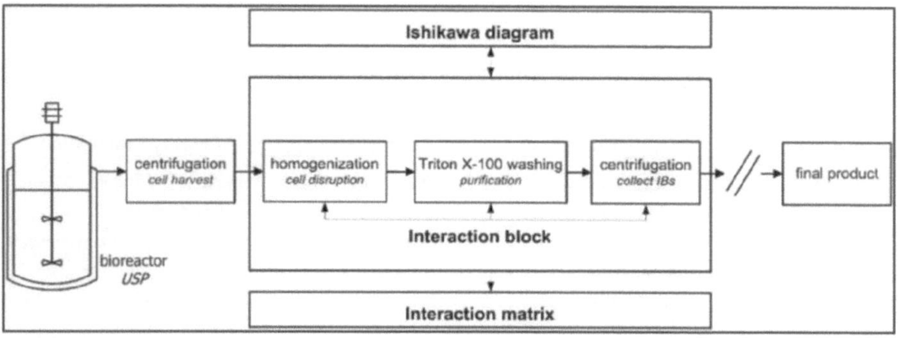

FIGURE 4.4 Integrated downstream processing workflow.

critical step sets the foundation for all subsequent purification operations and must be carefully optimized to maximize product recovery while effectively removing cellular contaminants. Modern primary recovery operations typically employ a combination of centrifugation and depth filtration technologies, with the specific combination depending on factors such as cell density, product location (intracellular or extracellular), and the physical properties of the target molecule.

Continuous disc-stack centrifugation has emerged as the industry standard for large-scale cell separation, offering advantages in terms of processing capacity and operational efficiency. The optimization of centrifugation parameters requires careful consideration of cell culture characteristics, including cell density, viability, and the presence of cell debris. Advanced centrifuge designs incorporate features for gentle product handling, minimizing shear forces that could potentially damage sensitive protein products or generate additional debris that could complicate subsequent purification steps.

Depth filtration serves as a complementary or alternative technology for primary clarification, particularly effective in removing fine particles and colloids that may pass through centrifugation. Modern depth filters incorporate multiple layers of filter media with gradually decreasing pore sizes, enabling the efficient removal of particles across a broad size range. The selection and sizing of depth filters must account for factors such as particle size distribution, filter capacity, and the potential for product binding to the filter media.

4.3.2.2 Concentration and Buffer Exchange

Following initial clarification, the product stream typically requires concentration and buffer conditioning to prepare for subsequent purification steps. Ultrafiltration and diafiltration operations play a crucial role in this phase, enabling both volume reduction and buffer exchange while maintaining product quality. The design of these operations must carefully consider factors such as protein concentration, solution conditions, and the potential for product aggregation or degradation (Table 4.4).

TABLE 4.4
Critical Parameters in Primary Recovery Operations

Parameter	Impact on Process Performance	Control Strategy	Quality Implications
Cell density	Affects separation efficiency	Online monitoring	Impacts clarification effectiveness
Filter capacity	Determines process economics	Load testing	Influences product recovery
Protein concentration	Affects subsequent operations	In-process testing	Product stability considerations
Buffer composition	Impacts product stability	Automated control systems	Chemical degradation risk

4.3.3 Chromatographic Purification Strategies

4.3.3.1 Platform Approaches in Protein Purification

The development of platform approaches to protein purification has revolutionized downstream processing in the biotechnology industry. These standardized purification schemes, particularly well established for monoclonal antibodies, provide a systematic framework for process development while allowing for product-specific optimization. The platform approach typically incorporates a series of orthogonal purification steps, each designed to target specific classes of impurities while maximizing product recovery.

Protein A affinity chromatography remains the cornerstone of many platform purification processes, particularly for monoclonal antibodies, offering exceptional selectivity and robust performance. The evolution of Protein A resins has led to improved capacity, stability, and cleaning resistance, enabling more efficient and economical purification processes. Modern Protein A media incorporate advanced base matrix materials and novel ligand designs that provide enhanced performance while maintaining the selective binding properties that make this technology so valuable.

4.3.3.2 Advanced Chromatography Technologies

The implementation of continuous chromatography systems represents a significant advancement in downstream processing technology. These systems, which enable the continuous processing of product streams through multiple columns operating in a coordinated sequence, offer advantages in terms of productivity, buffer consumption, and resin utilization. The successful implementation of continuous chromatography requires careful consideration of process parameters, control strategies, and validation approaches to ensure consistent performance and product quality.

4.3.4 Viral Clearance and Filtration Operations

4.3.4.1 Virus Removal Strategies

The ensuring of viral safety in biotechnology products represents a critical aspect of downstream processing, requiring a multilayered approach to virus removal and inactivation. Modern viral clearance strategies typically incorporate both dedicated virus removal steps and viral inactivation procedures integrated within the purification process. The selection and validation of viral clearance steps must consider both enveloped and nonenveloped viruses while maintaining product quality and process efficiency.

4.4 QUALITY CONTROL OF PROTEIN PRODUCTION

4.4.1 Introduction to Quality Control in Biotechnology Manufacturing

Quality control in biotechnology manufacturing represents a complex and multifaceted system of procedures, analyses, and verification steps that ensure the consistent production of safe and effective biological products. The implementation of robust quality control measures throughout the manufacturing process has become increasingly sophisticated, incorporating advanced analytical technologies, real-time monitoring systems, and comprehensive data analysis protocols. This evolution reflects the industry's commitment to maintaining the highest standards of product quality while meeting increasingly stringent regulatory requirements from authorities worldwide.

The foundation of effective quality control in biotechnology manufacturing rests upon the establishment of clear acceptance criteria for each critical quality attribute. These criteria must be based on thorough understanding of the relationship between product characteristics and therapeutic efficacy while also considering the limitations and capabilities of available analytical methods. Modern quality control laboratories must maintain a delicate balance between the need for rapid testing to support manufacturing operations and the requirement for thorough, validated analytical procedures that ensure accurate and reliable results.

Quality control systems in biotechnology extend far beyond simple product testing, encompassing environmental monitoring, raw material testing, in-process controls, and final product analysis. This comprehensive approach requires careful coordination between different operational units and the maintenance of sophisticated laboratory information management systems (LIMS) to ensure proper data handling and documentation. The integration of quality control data with manufacturing operations has become increasingly important as the industry moves toward real-time release testing and continuous manufacturing paradigms (Figure 4.5).

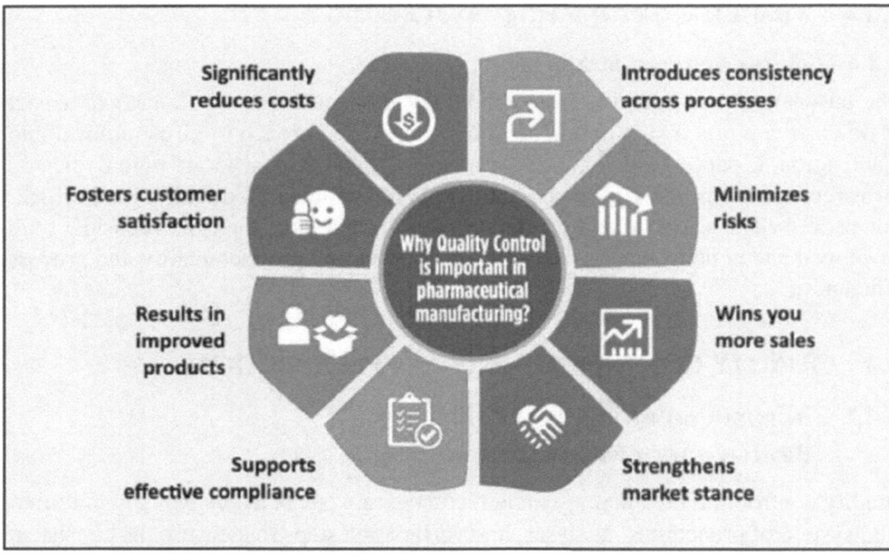

FIGURE 4.5 Comprehensive quality control framework in biotechnology manufacturing.

4.4.2 Analytical Methods and Technologies

Modern biotechnology quality control laboratories employ a diverse array of analytical methods and technologies to evaluate product quality attributes. These methods range from traditional biochemical assays to sophisticated instrumental analyses using state-of-the-art equipment. The selection and validation of appropriate analytical methods requires careful consideration of method specificity, accuracy, precision, and robustness while also accounting for practical constraints such as analysis time and sample requirements.

Protein characterization represents a particularly challenging aspect of biotechnology quality control, requiring multiple complementary analytical techniques to evaluate different product attributes. High-performance liquid chromatography (HPLC) remains a cornerstone of protein analysis, providing critical information about product purity, stability, and potential degradation products. The development of ultra-high-performance liquid chromatography (UHPLC) systems has further enhanced the capabilities of chromatographic analysis, enabling faster analysis times while maintaining or improving separation resolution.

Mass spectrometry has emerged as an increasingly important tool in biotechnology quality control, offering unprecedented capabilities for protein characterization and impurity analysis. The application of high-resolution mass spectrometry techniques enables detailed structural analysis of protein products, including sequence verification, post-translational modification assessment, and identification of process-related impurities. The integration of liquid chromatography with mass spectrometry (LC-MS) provides particularly powerful analytical capabilities, combining the separation efficiency of HPLC with the structural characterization capabilities of mass spectrometry.

Spectroscopic methods, including UV-visible spectrophotometry, fluorescence spectroscopy, and circular dichroism, provide essential information about protein concentration, conformational stability, and higher-order structure. These techniques are particularly valuable for routine quality control testing due to their nondestructive nature and relatively simple sample preparation requirements. The development of automated spectroscopic systems has enabled higher throughput analysis while maintaining analytical precision and accuracy.

Cell-based assays remain critical for evaluating the biological activity of protein products, despite their inherent variability and complexity. The development and validation of cell-based potency assays requires careful consideration of multiple factors, including cell line stability, assay conditions, and data analysis procedures. Modern approaches to cell-based testing increasingly incorporate automated systems and sophisticated data analysis tools to enhance assay reproducibility and throughput.

4.4.3 In-Process Controls and Real-Time Monitoring

In-process controls represent a critical component of biotechnology quality control systems, enabling early detection of potential issues and facilitating process optimization. The implementation of effective in-process controls requires careful selection of appropriate sampling points, analytical methods, and acceptance criteria. These controls must be designed to provide meaningful information about process performance while minimizing the impact on manufacturing operations.

The advent of process analytical technology (PAT) has revolutionized in-process control strategies in biotechnology manufacturing. PAT implementations typically incorporate multiple analytical technologies, including spectroscopic methods, automated sampling systems, and real-time data analysis capabilities. The integration of these technologies enables continuous monitoring of critical process parameters and quality attributes, supporting more sophisticated process control strategies.

Real-time monitoring systems generate large volumes of data that must be effectively managed and analyzed to support manufacturing decisions. The development of appropriate data management systems and analytical algorithms represents a significant challenge in implementing real-time monitoring strategies. Modern systems increasingly incorporate advanced data analysis techniques, including multivariate statistical analysis and machine learning algorithms, to extract meaningful information from complex datasets (Table 4.5).

4.4.4 Environmental Monitoring and Contamination Control

Environmental monitoring programs in biotechnology manufacturing facilities must be comprehensive and well-documented, encompassing air quality, surface contamination, water systems, and personnel monitoring. The design of effective environmental monitoring programs requires careful consideration of facility layout,

TABLE 4.5
Critical In-Process Control Parameters and Monitoring Strategies

Parameter Category	Analytical Methods	Monitoring Frequency	Control Strategy
Cell culture	Cell density, viability	Continuous/daily	Automated PAT systems
Protein expression	Titer, activity	Batch/daily	Offline analysis
Product quality	Purity, stability	Critical steps	Multiple methods
Process environment	pH, temperature, gases	Continuous	Real-time sensors
Contamination	Bioburden, endotoxin	Defined intervals	Rapid methods

process requirements, and regulatory expectations. Modern programs increasingly incorporate automated sampling systems and rapid microbiological methods to enhance monitoring efficiency and effectiveness.

The control of microbial contamination represents a particular challenge in biotechnology manufacturing, requiring implementation of robust aseptic techniques and careful monitoring of critical process steps. The development of rapid microbiological methods has significantly enhanced contamination control capabilities, enabling faster detection of potential contamination events and more timely implementation of corrective actions. These methods must be carefully validated to ensure they provide equivalent or superior performance compared to traditional microbiological testing approaches.

Water system monitoring represents another critical aspect of environmental control in biotechnology manufacturing. The production and monitoring of Water for Injection (WFI) and other grades of pharmaceutical water requires sophisticated purification systems and comprehensive monitoring programs. Modern water system designs increasingly incorporate continuous monitoring capabilities for critical parameters such as conductivity, total organic carbon, and bioburden levels.

4.4.5 Raw Material Testing and Supplier Quality Management

The quality control of raw materials used in biotechnology manufacturing requires implementation of comprehensive testing programs and effective supplier management systems. Raw material testing programs must be designed to evaluate all critical quality attributes while considering practical constraints such as testing time and sample availability. The development of appropriate specifications and testing strategies requires careful consideration of material criticality, supplier capabilities, and regulatory requirements.

Supplier quality management systems have become increasingly sophisticated, incorporating detailed supplier qualification procedures, ongoing performance monitoring, and regular quality audits. The establishment of effective supplier partnerships requires clear communication of quality expectations and development of appropriate quality agreements. Modern supplier management systems increasingly incorporate electronic data exchange capabilities to facilitate sharing of quality control data and certificates of analysis.

The testing of complex biological raw materials presents particular challenges in biotechnology manufacturing, requiring specialized analytical methods and careful consideration of material variability. The development of appropriate testing strategies for materials such as growth media components, cell culture supplements, and chromatography resins requires thorough understanding of material characteristics and potential impact on process performance. Modern approaches increasingly incorporate risk-based testing strategies to optimize resource utilization while maintaining appropriate quality oversight.

4.4.6 Data Management and Quality Systems

Quality control data management in biotechnology manufacturing requires sophisticated systems capable of handling large volumes of complex analytical data while maintaining data integrity and regulatory compliance. Laboratory information management systems (LIMS) represent a critical component of modern quality control operations, providing capabilities for sample tracking, data capture, and results reporting. The implementation of effective LIMS solutions requires careful consideration of workflow requirements, data integrity controls, and integration with other manufacturing systems.

The maintenance of data integrity in quality control operations has become increasingly important as regulatory authorities focus more attention on this aspect of pharmaceutical manufacturing. Modern quality control systems must incorporate appropriate controls to ensure data security, traceability, and reliability. These controls typically include features such as audit trails, electronic signatures, and automated data backup systems.

Quality control data analysis and trending represent critical activities in modern biotechnology manufacturing, enabling early detection of potential issues and supporting continuous improvement initiatives. The development of appropriate statistical tools and visualization capabilities helps quality control personnel identify trends and patterns in analytical data. Modern systems increasingly incorporate advanced analytics capabilities, including statistical process control charts and multivariate analysis tools.

4.5 FINAL FILL AND FINISH OF PRODUCT

4.5.1 Advanced Formulation Strategies for Biological Products

The final formulation of biological products represents a critical phase in the manufacturing process, demanding sophisticated approaches to ensure product stability, efficacy, and safety throughout the product life cycle. Modern formulation strategies extend far beyond simple buffer preparation, incorporating advanced stabilization technologies and innovative delivery systems that enhance product performance and patient compliance. The complexity of protein-based therapeutics necessitates careful consideration of multiple factors during formulation development, including protein stability, aggregation tendencies, chemical degradation pathways, and interaction with container closure systems. Contemporary approaches to formulation development increasingly utilize high-throughput screening methodologies combined with advanced analytical techniques to identify optimal formulation conditions that maintain product quality throughout the intended shelf life.

The selection of appropriate excipients plays a fundamental role in formulation development, requiring extensive evaluation of compatibility, stability, and regulatory acceptance. Modern formulation strategies often incorporate novel stabilizing agents and innovative buffer systems that provide enhanced protection against various degradation pathways. The development of excipient strategies must consider not only immediate stability requirements but also long-term storage conditions, transportation requirements, and potential interaction with delivery devices. This comprehensive approach to excipient selection becomes particularly critical for biological products intended for various global markets, where storage and distribution conditions may vary significantly.

The implementation of robust formulation processes requires sophisticated mixing and preparation strategies that ensure consistent product quality while maintaining sterility throughout the manufacturing process. Modern facilities utilize advanced mixing systems and preparation protocols that enable precise control over critical parameters such as pH, osmolality, and protein concentration. The development of mixing strategies must consider not only immediate process requirements but also scale-up implications and manufacturing efficiency objectives. This attention to process development becomes particularly important when implementing new formulation technologies or novel delivery systems.

Container closure system selection represents another critical aspect of product formulation, requiring careful evaluation of materials compatibility, container integrity, and user requirements. Modern packaging solutions incorporate innovative materials and designs that enhance product protection while improving ease of use for healthcare providers and patients. The development of container closure strategies must consider not only immediate packaging requirements but also long-term stability implications and potential extractables and leachables concerns. This comprehensive approach to packaging development becomes particularly important for products intended for use with specific delivery devices or administration systems.

The implementation of stability assessment programs represents a fundamental aspect of formulation development, requiring sophisticated approaches to accelerated testing and real-time monitoring. Modern stability programs utilize advanced analytical techniques that enable comprehensive characterization of product attributes throughout the storage period. The development of stability assessment strategies must consider not only regulatory requirements but also real-world storage and handling conditions. This comprehensive approach to stability assessment becomes particularly critical for products intended for global distribution, where varying environmental conditions may impact product quality (Table 4.6).

TABLE 4.6
Critical Parameters in Final Formulation Development

Parameter	Impact on Product	Control Strategy	Quality Implications
pH stability	Chemical degradation	Buffer selection	Product integrity
Osmolality	Physical stability	Excipient optimization	Solution properties
Protein concentration	Viscosity/stability	Process control	Delivery requirements
Container compatibility	Product integrity	Material selection	Storage stability
Sterility assurance	Patient safety	Process design	Regulatory compliance

4.5.2 Aseptic Processing and Sterility Assurance

The implementation of robust aseptic processing systems represents a cornerstone of modern biotechnology manufacturing, requiring sophisticated approaches to contamination control and sterility assurance. Contemporary aseptic processing facilities incorporate advanced cleanroom technologies, automated handling systems, and comprehensive environmental monitoring programs to ensure product sterility throughout the manufacturing process. The complexity of modern aseptic processing extends beyond basic sterility maintenance to encompass considerations of process efficiency, operator safety, and regulatory compliance. The development of effective aseptic processing strategies requires careful integration of facility design, equipment selection, and operational procedures to create a comprehensive contamination control program.

Environmental monitoring programs play a crucial role in aseptic processing, utilizing sophisticated sampling and analysis methods to verify the maintenance of appropriate environmental conditions. Modern monitoring systems incorporate continuous particle counting, viable monitoring, and advanced data management capabilities that enable real-time assessment of cleanroom conditions. The development of monitoring strategies must consider not only immediate regulatory requirements but also trend analysis capabilities and rapid response protocols for addressing potential contamination events. This comprehensive approach to environmental monitoring becomes particularly critical during aseptic operations, where maintaining appropriate environmental conditions directly impacts product quality and patient safety.

The implementation of advanced barrier systems represents another critical aspect of modern aseptic processing, with restricted access barrier systems (RABS) and isolator technology becoming increasingly prevalent in manufacturing operations. These sophisticated containment systems provide enhanced protection against contamination while enabling efficient processing operations through carefully designed transfer systems and automated handling capabilities. The development of barrier system strategies must consider not only immediate containment requirements but also operational efficiency and maintenance requirements. This balance becomes particularly important in high-throughput manufacturing environments, where system reliability and operational flexibility directly impact production efficiency.

Personnel training and qualification programs represent a fundamental component of successful aseptic processing operations, requiring comprehensive approaches to knowledge development and skill assessment. Modern training programs utilize advanced simulation technologies and practical exercises to develop proper aseptic technique and contamination control awareness. The implementation of training programs must consider not only immediate operational requirements but also ongoing assessment and requalification needs. This attention to personnel development becomes particularly critical in GMP manufacturing environments, where operator performance directly impacts product quality and compliance status.

Process simulation studies, commonly known as media fills, play an essential role in validating aseptic processing capabilities and identifying potential contamination risks. Modern approaches to process simulation incorporate comprehensive challenge conditions and sophisticated analysis methods to verify system capability. The development of simulation programs must consider not only regulatory requirements but also worst-case processing scenarios and potential failure modes. This thorough approach to process validation becomes particularly important when implementing new technologies or modified operating procedures (Figure 4.6).

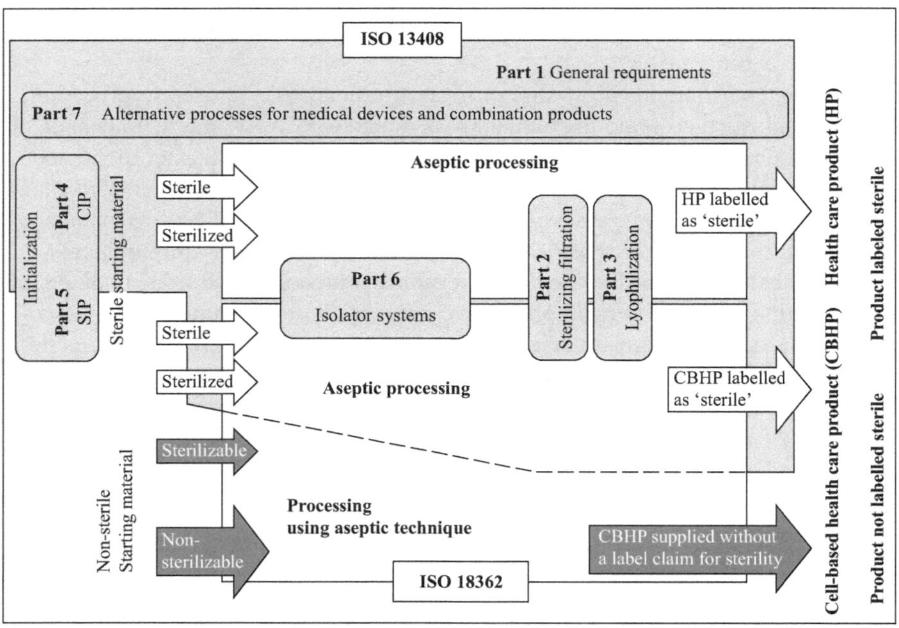

FIGURE 4.6 Advanced aseptic processing systems.

4.5.3 Automated Fill–Finish Systems

The integration of automated fill-finish systems represents a significant advancement in modern biotechnology manufacturing, enabling precise control over critical filling operations while minimizing contamination risks and human intervention. Contemporary filling systems incorporate sophisticated robotics, advanced vision systems, and integrated weight checking capabilities that ensure consistent product delivery while maintaining sterile conditions. The complexity of modern fill–finish operations extends beyond basic filling accuracy to encompass considerations of container handling, environmental protection, and process monitoring. The development of automated filling strategies requires careful consideration of product characteristics, container requirements, and operational efficiency objectives.

Machine vision systems play an increasingly important role in automated fill-finish operations, providing comprehensive inspection capabilities for both containers and filling processes. Modern vision systems utilize advanced imaging technologies and sophisticated analysis algorithms that enable real-time detection of defects and process deviations. The implementation of vision system strategies must consider not only immediate inspection requirements but also data management and trend analysis capabilities. This comprehensive approach to quality monitoring becomes particularly critical in high-speed filling operations, where rapid detection of potential issues directly impacts product quality and yield.

The development of robust container handling systems represents another critical aspect of automated fill–finish operations, requiring sophisticated approaches to maintaining container integrity while ensuring efficient processing. Modern handling systems incorporate advanced materials and designs that minimize particle generation while enabling smooth container movement throughout the filling process. The implementation of handling strategies must consider not only immediate operational requirements but also flexibility for different container formats and sizes. This attention to system design becomes particularly important when processing multiple product presentations on the same filling line (Table 4.7).

Process control systems in automated fill-finish operations have evolved significantly, incorporating sophisticated feedback mechanisms and adaptive control capabilities that maintain precise filling parameters throughout production runs. Modern

TABLE 4.7
Critical Parameters in Automated Fill–Finish Operations

Parameter	Control Strategy	Monitoring Approach	Quality Impact
Fill weight	Precision pumps	In-line weighing	Product uniformity
Container integrity	Vision systems	Automated inspection	Package quality
Environmental control	RABS/isolators	Continuous monitoring	Sterility assurance
Process speed	Automated controls	Performance tracking	Production efficiency
Component quality	Incoming inspection	Statistical sampling	Final product quality

control systems utilize advanced algorithms and real-time monitoring capabilities to adjust filling parameters based on process performance and environmental conditions. The development of control strategies must consider not only immediate accuracy requirements but also long-term stability and maintenance implications. This comprehensive approach to process control becomes particularly critical when implementing new filling technologies or modified operating procedures.

The integration of automated cleaning and sterilization systems represents another fundamental aspect of modern fill-finish operations, requiring sophisticated approaches to ensuring equipment cleanliness and sterility. Contemporary cleaning systems incorporate automated clean-in-place (CIP) and sterilize-in-place (SIP) capabilities that enable efficient turnaround between production runs while maintaining validated cleaning processes. The implementation of cleaning strategies must consider not only immediate sterilization requirements but also material compatibility and process validation implications. This attention to cleaning system design becomes particularly important in multiproduct manufacturing environments, where cross-contamination prevention is critical.

4.5.4 Quality Control in Fill–Finish Operations

The implementation of comprehensive quality control strategies in fill–finish operations represents a critical aspect of modern biotechnology manufacturing, requiring sophisticated approaches to product testing, process monitoring, and data management. Contemporary quality control programs integrate advanced analytical technologies with real-time monitoring systems to ensure consistent product quality throughout the filling process. The complexity of modern quality control extends beyond basic product testing to encompass environmental monitoring, in-process controls, and final product evaluation. The development of effective quality control strategies requires careful consideration of regulatory requirements, process capabilities, and product specifications to create a robust quality assurance program.

In-process testing plays a fundamental role in quality control, utilizing advanced analytical methods to verify product quality parameters throughout the filling operation. Modern testing approaches incorporate automated sampling systems and rapid analysis techniques that enable quick detection of potential quality issues. The implementation of in-process testing strategies must consider not only immediate analytical requirements but also sample handling procedures and data management capabilities. This comprehensive approach to quality monitoring becomes particularly critical during high-speed filling operations, where rapid quality assessment directly impacts production efficiency and product yield.

The integration of advanced particle monitoring systems represents another crucial component of quality control in fill-finish operations. Contemporary monitoring systems utilize sophisticated detection technologies and automated analysis capabilities to identify potential particulate contamination in both products and processing environments. The development of particle monitoring strategies must consider not only regulatory requirements but also trend analysis capabilities and response protocols for addressing detected issues. This attention to particulate control becomes

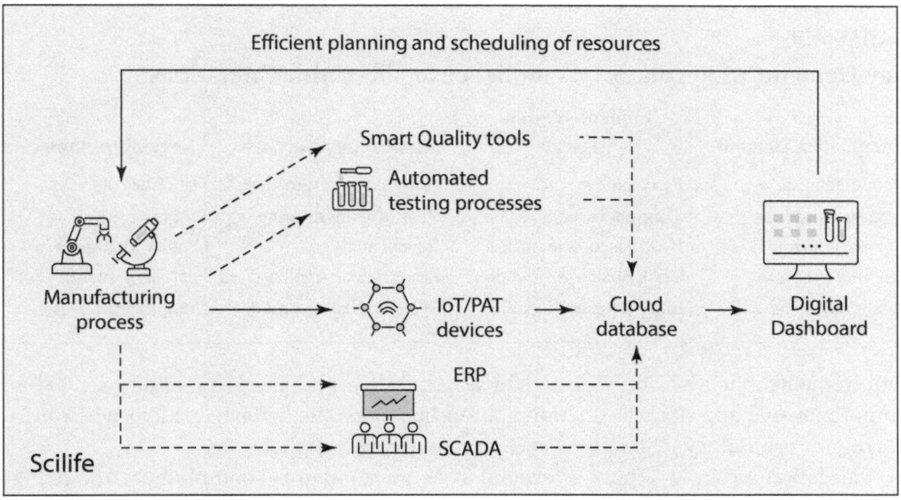

FIGURE 4.7 Integrated quality control systems in fill–finish operations.

especially important when processing protein-based products, where protein aggregation can impact product quality and safety.

Statistical process control methodologies have evolved significantly in modern fill-finish operations, incorporating advanced analysis techniques and real-time trending capabilities that enable proactive quality management. The implementation of statistical control strategies must consider not only immediate process monitoring requirements but also long-term trend analysis and continuous improvement objectives. This comprehensive approach to process control becomes particularly important when establishing process capability indices and determining appropriate control limits for critical quality attributes (Figure 4.7).

4.5.5 Regulatory Compliance Strategies

The development and implementation of comprehensive regulatory compliance strategies represents a fundamental requirement in modern fill-finish operations, necessitating sophisticated approaches to documentation, validation, and quality system management. Contemporary compliance programs integrate advanced documentation systems with robust quality management processes to ensure adherence to global regulatory requirements. The complexity of modern regulatory compliance extends beyond basic GMP adherence to encompass considerations of data integrity, electronic systems validation, and change control management.

Documentation systems in modern fill-finish operations have evolved significantly, incorporating electronic batch record systems and automated data capture capabilities that ensure complete and accurate process documentation. The implementation of documentation strategies must consider not only immediate regulatory

TABLE 4.8
Key Elements of Regulatory Compliance in Fill–Finish Operations

Compliance Element	Implementation Strategy	Monitoring Approach	Regulatory Impact
Documentation	Electronic systems	Automated verification	Data integrity
Process validation	Continuous verification	Real-time monitoring	Process control
Change control	Risk-based assessment	Impact analysis	Quality assurance
Training programs	Competency verification	Performance tracking	Personnel qualification
Quality systems	Integration approach	System effectiveness	Compliance status

requirements but also long-term data accessibility and integrity assurance. This comprehensive approach to documentation becomes particularly critical when preparing for regulatory inspections and product license applications.

Validation programs play a crucial role in regulatory compliance, requiring sophisticated approaches to process validation, cleaning validation, and computer system validation. Modern validation strategies incorporate risk-based approaches and continuous verification methodologies that ensure ongoing process control and product quality. The development of validation programs must consider not only immediate compliance requirements but also life cycle management implications and revalidation needs. This attention to validation strategy becomes especially important when implementing new technologies or modified operating procedures (Table 4.8).

4.5.6 Emerging Technologies and Future Directions

The landscape of fill-finish operations continues to evolve with the introduction of innovative technologies and advanced processing solutions. Contemporary manufacturing facilities increasingly incorporate novel container closure systems, advanced robotics, and artificial intelligence-driven control systems that enhance operational efficiency and product quality. The integration of these emerging technologies requires careful consideration of validation requirements, operator training needs, and regulatory acceptance. The development of implementation strategies must balance innovation objectives with practical considerations of operational reliability and regulatory compliance.

4.6 CASE STUDIES: THERAPEUTIC PROTEINS, MONOCLONAL ANTIBODIES, HUMAN VACCINES

4.6.1 Therapeutic Protein Production: Recombinant Insulin Manufacturing

The production of recombinant insulin represents one of the most significant achievements in biotechnology manufacturing, serving as a cornerstone example of successful large-scale protein production. Modern insulin manufacturing facilities integrate sophisticated upstream and downstream processing techniques to ensure consistent product quality while maintaining high production efficiency. The evolution of insulin production processes demonstrates the industry's capability to optimize complex biological systems for commercial-scale manufacturing. Contemporary facilities utilize advanced cell culture systems combined with highly efficient purification processes to achieve unprecedented levels of product purity and process yield.

The implementation of process analytical technology (PAT) in insulin manufacturing has revolutionized production monitoring and control strategies. Advanced sensor systems enable real-time assessment of critical quality attributes throughout the manufacturing process, allowing for immediate process adjustments to maintain optimal conditions. The integration of multiple monitoring systems creates comprehensive data streams that provide detailed insights into process performance and product quality. This level of process understanding has enabled manufacturers to implement sophisticated control strategies that ensure consistent product quality while maximizing production efficiency.

Quality control strategies in insulin manufacturing have evolved significantly, incorporating advanced analytical techniques that enable detailed characterization of product quality attributes throughout the production process. Modern facilities utilize multiple orthogonal testing methods to ensure comprehensive product characterization, from basic physicochemical properties to complex biological activity assessments. The implementation of these analytical strategies requires careful coordination between production and quality control teams to ensure timely decision-making while maintaining compliance with regulatory requirements (Figure 4.8).

The development of robust cleaning and sanitization procedures represents another critical aspect of insulin manufacturing, requiring careful validation to ensure consistent process performance and product quality. Modern facilities implement sophisticated clean-in-place (CIP) and sterilize-in-place (SIP) systems that enable efficient equipment turnover while maintaining strict control over potential contamination risks. The validation of these procedures requires extensive documentation and testing to demonstrate consistent effectiveness across multiple manufacturing cycles.

Process optimization in insulin manufacturing continues to evolve, incorporating new technologies and approaches to enhance production efficiency and product quality. Contemporary facilities increasingly utilize single-use technologies in certain process steps, offering advantages in terms of operational flexibility and contamination risk reduction. The implementation of these technologies requires careful evaluation of material compatibility and extractables/leachables profiles to ensure product quality is not compromised. This ongoing evolution demonstrates the industry's commitment to continuous improvement while maintaining strict compliance with regulatory requirements.

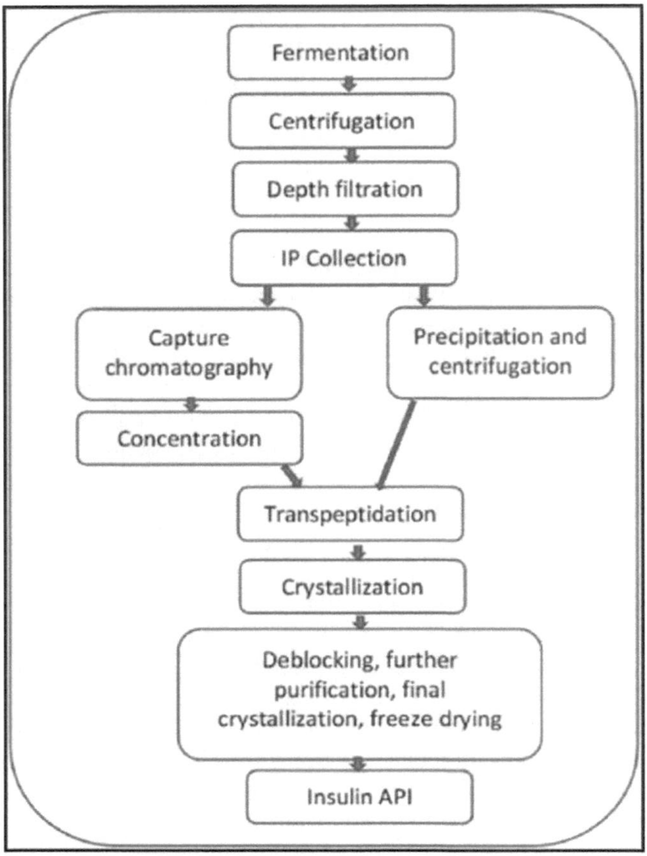

FIGURE 4.8 Integrated insulin manufacturing process flow.

4.6.2 Monoclonal Antibody Production: Process Development and Scale-Up

The production of monoclonal antibodies represents one of the most complex challenges in modern biotechnology manufacturing, requiring sophisticated process development approaches and careful scale-up strategies. Contemporary facilities utilize advanced cell culture systems combined with highly specific purification processes to achieve exceptional product quality while maintaining commercial viability. The development of robust manufacturing processes requires careful consideration of multiple factors, from cell line selection through final product formulation, each critical to ensuring consistent product quality and process performance (Table 4.9).

The implementation of platform technologies in monoclonal antibody production has significantly enhanced process development efficiency while maintaining high

TABLE 4.9
Critical Parameters in Monoclonal Antibody Manufacturing

Process Stage	Critical Parameters	Control Strategy	Quality Impact
Cell culture	Viability, metabolism	Feed optimization	Product quality
Harvesting	Cell removal, product stability	Filtration design	Process yield
Purification	Impurity clearance	Chromatography sequence	Final purity
Formulation	Stability, concentration	Buffer optimization	Product efficacy
Analytics	Product characterization	Method validation	Release criteria

product quality standards. Modern facilities utilize standardized process templates that enable rapid process development while incorporating specific optimizations for individual products. These platform approaches extend beyond basic unit operations to encompass comprehensive control strategies and analytical methods, enabling efficient technology transfer and scale-up activities.

Cell culture optimization represents a particularly critical aspect of monoclonal antibody production, requiring sophisticated approaches to media development and process control. Contemporary manufacturing processes utilize advanced feeding strategies that maintain optimal conditions throughout extended cultivation periods. The development of these strategies requires careful balance of nutrient provision, metabolic waste management, and product quality considerations. This optimization process increasingly incorporates artificial intelligence approaches to identify optimal operating conditions while maintaining process robustness.

4.6.3 Human Vaccine Production: Advanced Manufacturing Approaches

The manufacturing of human vaccines represents one of the most challenging and critical applications in modern biotechnology, demanding exceptional attention to process control, quality assurance, and regulatory compliance. Contemporary vaccine production facilities integrate cutting-edge technologies with well-established manufacturing principles to ensure consistent product quality while maintaining high production efficiency. The complexity of vaccine manufacturing extends beyond basic production challenges to encompass considerations of product stability, cold chain management, and global distribution requirements. Modern facilities must maintain unprecedented levels of process control while simultaneously preparing for rapid scale-up capabilities in response to global health emergencies.

The implementation of cell culture-based vaccine production systems has revolutionized the industry, offering significant advantages over traditional egg-based manufacturing approaches. These advanced production platforms enable more rapid process development and scale-up while providing greater control over critical quality attributes. The selection of appropriate cell lines and culture conditions requires careful consideration of multiple factors, from growth characteristics and productivity to genetic stability and safety profiles. Contemporary facilities increasingly utilize perfusion-based cultivation systems that enable continuous production while maintaining optimal conditions for virus propagation or protein expression.

Quality control strategies in vaccine manufacturing have evolved significantly, incorporating sophisticated analytical methods that enable comprehensive characterization of both in-process materials and final products. Modern facilities employ multiple orthogonal testing approaches to ensure complete understanding of product quality attributes, from basic physicochemical properties to complex immunological assessments. The implementation of rapid testing methods has become particularly critical in vaccine manufacturing, enabling faster batch release decisions while maintaining strict compliance with regulatory requirements. The development of these analytical strategies requires careful validation to ensure consistent performance across different product types and manufacturing scales (Figure 4.9).

The validation of cleaning and sanitization procedures represents another crucial aspect of vaccine manufacturing, requiring rigorous documentation and testing to ensure consistent effectiveness. Modern facilities implement sophisticated clean-in-place (CIP) and sterilize-in-place (SIP) systems that enable efficient equipment turnover while maintaining strict control over potential contamination risks. The development of these procedures must consider not only immediate cleaning effectiveness but also long-term impact on equipment performance and product quality. The validation of cleaning procedures becomes particularly critical in multiproduct facilities, where prevention of cross-contamination represents a primary concern.

Process optimization in vaccine manufacturing continues to evolve, incorporating new technologies and approaches to enhance production efficiency and product quality. Contemporary facilities increasingly utilize single-use technologies in certain process steps, offering advantages in terms of operational flexibility and contamination risk reduction. The implementation of these technologies requires careful evaluation of material compatibility and extractables/leachables profiles to ensure product quality is not compromised. This ongoing evolution demonstrates the industry's commitment to continuous improvement while maintaining strict compliance with regulatory requirements.

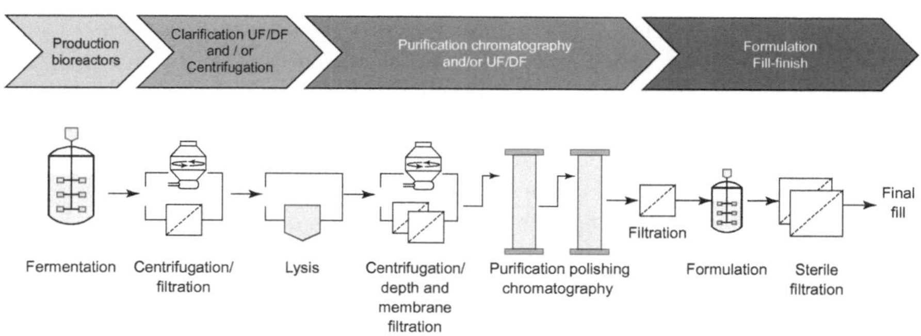

FIGURE 4.9 Modern vaccine manufacturing process integration.

4.6.4 Process Validation Case Studies

The validation of biotechnology manufacturing processes represents a critical element in ensuring consistent product quality and regulatory compliance. Modern validation approaches incorporate quality by design (QbD) principles that enable comprehensive understanding of process capabilities and limitations. Contemporary facilities implement sophisticated validation strategies that extend beyond basic process qualification to encompass continuous process verification and ongoing monitoring programs. The development of effective validation approaches requires careful consideration of multiple factors, from analytical method capabilities to process control strategies and equipment performance characteristics (Table 4.10).

The implementation of process characterization studies represents a fundamental aspect of validation activities, requiring sophisticated experimental design approaches to identify and understand critical process parameters. Modern facilities utilize advanced statistical tools combined with process analytical technology (PAT) initiatives to develop comprehensive understanding of parameter interactions and their impact on product quality. The development of these characterization studies requires careful planning to ensure appropriate coverage of potential operating conditions while maintaining practical feasibility of execution.

Continued process verification programs have become increasingly sophisticated, incorporating real-time monitoring capabilities that enable immediate detection of process deviations or trends. Contemporary facilities implement comprehensive monitoring strategies that combine traditional in-process testing with advanced PAT initiatives to provide complete process understanding. The development of these verification programs requires careful consideration of statistical tools and trending approaches to ensure meaningful analysis of collected data. The implementation of effective trending programs becomes particularly critical in maintaining process validation status over extended production periods.

The validation of cleaning procedures represents another critical aspect of biotechnology manufacturing, requiring comprehensive understanding of potential residues and appropriate cleaning mechanisms. Modern facilities implement sophisticated analytical methods capable of detecting trace levels of product residues or cleaning agents. The development of cleaning validation protocols must consider not

TABLE 4.10
Process Validation Critical Elements

Validation Aspect	Key Considerations	Implementation Strategy	Quality Impact
Process design	Parameter ranges	DoE studies	Product consistency
Equipment qualification	Performance criteria	Testing protocols	Process reliability
Cleaning validation	Residue limits	Analytical methods	Cross-contamination
Method validation	Specificity, accuracy	Method transfer	Result reliability
Computer systems	Data integrity	System testing	Documentation quality

only immediate cleaning effectiveness but also long-term impact on equipment performance and product quality. The implementation of effective cleaning validation programs becomes particularly critical in multiproduct facilities where prevention of cross-contamination represents a primary concern.

Computer system validation has evolved significantly with the increasing implementation of automated control systems and electronic documentation platforms. Contemporary facilities must validate complex networks of interconnected systems while maintaining data integrity and regulatory compliance. The development of computer system validation protocols requires careful consideration of both technical capabilities and regulatory requirements. The implementation of effective validation strategies becomes particularly critical when implementing new software systems or upgrading existing platforms.

4.6.5 Quality Control Strategy Implementation

The implementation of comprehensive quality control strategies represents a fundamental aspect of successful biotechnology manufacturing. Modern quality control approaches extend far beyond traditional final product testing to encompass sophisticated in-process controls and real-time monitoring systems. Contemporary facilities integrate multiple analytical technologies to ensure complete characterization of product quality attributes throughout the manufacturing process. The development of these strategies requires careful consideration of analytical method capabilities, sampling approaches, and data management systems to ensure effective decision-making while maintaining regulatory compliance.

Advanced analytical methods have revolutionized quality control capabilities in biotechnology manufacturing, enabling unprecedented levels of product characterization and process understanding. Modern facilities employ multiple orthogonal testing approaches to ensure comprehensive assessment of product quality attributes, from basic physicochemical properties to complex biological activity measurements. The implementation of high-throughput analytical platforms has significantly enhanced testing efficiency while maintaining high standards of accuracy and precision. The development of these analytical capabilities requires careful method validation and ongoing performance monitoring to ensure consistent reliability across different product types and manufacturing scales.

Real-time release testing strategies have emerged as a critical element of modern quality control programs, enabling faster product release while maintaining strict compliance with regulatory requirements. Contemporary facilities implement sophisticated PAT initiatives that enable continuous monitoring of critical quality attributes throughout the manufacturing process. The development of real-time release strategies requires careful correlation between in-process measurements and final product specifications. The implementation of these approaches necessitates robust analytical methods and comprehensive understanding of process–product relationships (Figure 4.10).

Stability testing programs represent another critical aspect of quality control strategy implementation, requiring careful consideration of storage conditions, testing

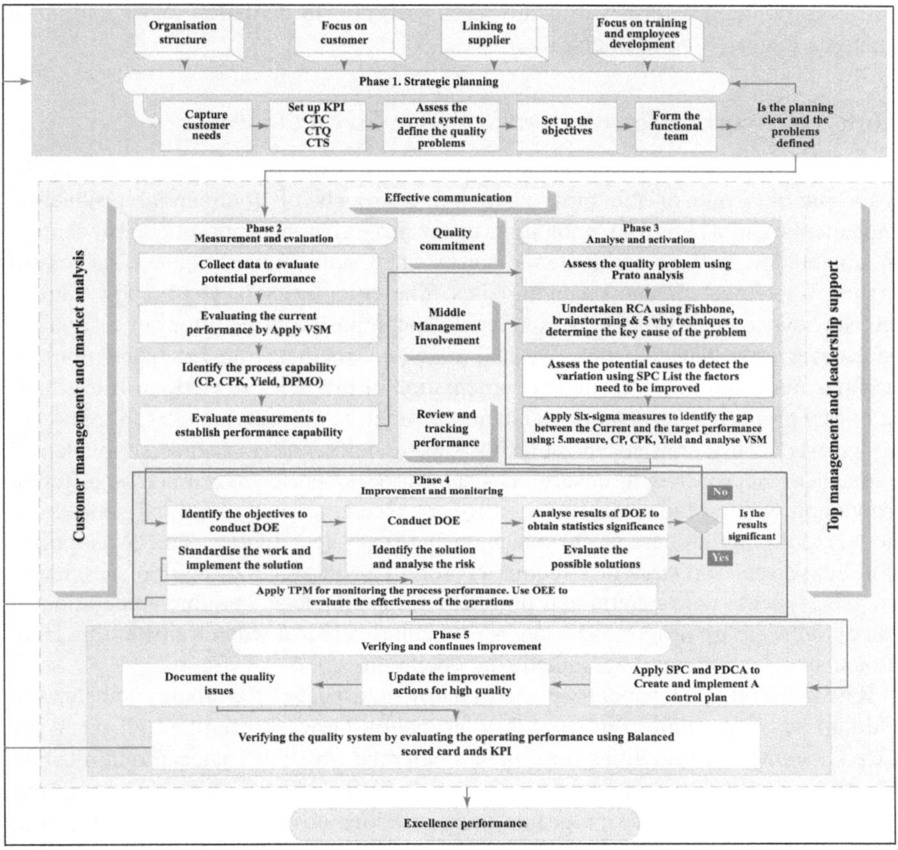

FIGURE 4.10 Integrated quality control strategy framework.

intervals, and analytical method selection. Modern facilities implement comprehensive stability assessment protocols that evaluate product quality attributes under various environmental conditions. The development of stability programs must consider not only regulatory requirements but also practical aspects of product distribution and storage. The implementation of effective stability assessment strategies becomes particularly critical for biological products with complex stability profiles or special storage requirements.

Data management systems play an increasingly important role in quality control strategy implementation, requiring sophisticated approaches to data collection, analysis, and trending. Contemporary facilities utilize integrated laboratory information management systems (LIMS) that enable efficient data handling while maintaining complete traceability and data integrity. The development of these systems must consider not only immediate testing requirements but also long-term trending analysis and regulatory reporting needs. The implementation of electronic data management platforms requires careful validation to ensure reliable operation and compliance with data integrity requirements.

4.6.6 Technology Transfer and Scale-Up Considerations

The successful transfer of biotechnology manufacturing processes between facilities represents one of the most challenging aspects of commercial production implementation. Modern technology transfer programs incorporate comprehensive documentation systems and systematic assessment approaches to ensure successful process implementation at receiving sites. The complexity of technology transfer extends beyond basic process parameters to encompass considerations of equipment differences, raw material sourcing, and analytical method transfer. Contemporary facilities must maintain detailed documentation of process knowledge while simultaneously preparing for potential adaptations required at receiving sites (Table 4.11).

Scale-up considerations in biotechnology manufacturing require sophisticated engineering approaches to ensure consistent process performance across different production scales. Modern facilities utilize advanced modeling tools and scale-down models to predict process performance during scale-up activities. The development of effective scale-up strategies requires careful consideration of equipment characteristics, process parameters, and product quality impacts. The implementation of successful scale-up programs becomes particularly critical when transitioning from pilot-scale to commercial manufacturing operations.

Raw material management represents a critical aspect of technology transfer and scale-up activities, requiring careful coordination between sending and receiving sites. Contemporary facilities implement comprehensive raw material qualification programs that ensure consistent quality across different manufacturing locations. The development of these programs must consider not only immediate material specifications but also long-term supply chain reliability and regulatory compliance requirements. The implementation of effective raw material management strategies becomes particularly critical when establishing global manufacturing networks.

Training program development represents another crucial element of successful technology transfer, requiring comprehensive approaches to knowledge transfer and personnel qualification. Modern facilities implement sophisticated training

TABLE 4.11
Critical Elements in Technology Transfer

Transfer Aspect	Key Considerations	Implementation Strategy	Success Metrics
Process knowledge	Critical parameters	Documentation systems	Process understanding
Analytical methods	Method transfer	Comparison studies	Method equivalency
Raw materials	Supply chain	Qualification programs	Material consistency
Equipment design	Design differences	Engineering studies	Performance matching
Training programs	Personnel qualification	Knowledge transfer	Operational readiness

systems that combine theoretical understanding with practical experience to ensure operational readiness. The development of these programs must consider not only immediate operational requirements but also long-term maintenance of process knowledge and operational excellence. The implementation of effective training strategies becomes particularly critical when establishing new manufacturing operations or implementing significant process changes.

Process characterization studies play a fundamental role in technology transfer and scale-up activities, requiring sophisticated experimental approaches to develop comprehensive process understanding. Contemporary facilities utilize advanced statistical tools and process analytical technology to identify and control critical process parameters. The development of these characterization studies must consider not only immediate process performance but also long-term manufacturing reliability and product quality consistency. The implementation of effective characterization programs becomes particularly critical when establishing robust manufacturing processes at new facilities.

4.7 ARTIFICIAL INTELLIGENCE FOR STANDARD OPERATING PROCEDURES OPTIMIZATION

4.7.1 Integration of Artificial Intelligence in Modern Biotechnology Operations

The integration of artificial intelligence into biotechnology manufacturing represents a paradigm shift in how standard operating procedures are developed, implemented, and optimized. Modern biotechnology facilities increasingly leverage AI technologies to enhance process understanding, improve operational efficiency, and maintain consistent product quality across complex manufacturing operations. This transformation extends beyond simple automation to encompass sophisticated machine learning algorithms that can analyze vast amounts of historical process data, identify subtle patterns and relationships, and generate actionable insights for process improvement. The implementation of AI-driven systems requires careful consideration of existing regulatory frameworks, validation requirements, and the need to maintain complete process transparency and control.

The application of artificial intelligence in SOP optimization begins with comprehensive data integration from multiple sources throughout the manufacturing process. Modern facilities generate enormous quantities of process data from various instruments, sensors, and control systems, creating a rich foundation for AI-driven analysis. The challenge lies not only in collecting and organizing this data but in developing sophisticated algorithms that can extract meaningful patterns and relationships while accounting for the inherent variability of biological processes. This requires the development of robust data management systems that can handle both structured and unstructured data while maintaining complete data integrity and traceability.

The evolution of machine learning algorithms specifically designed for biotechnology applications has enabled more sophisticated approaches to process optimization.

These algorithms can simultaneously consider multiple process parameters, quality attributes, and operational constraints to identify optimal operating conditions while maintaining product quality and process consistency. The development of these specialized algorithms requires deep understanding of both artificial intelligence principles and biotechnology manufacturing processes, necessitating close collaboration between data scientists and process experts. This interdisciplinary approach ensures that AI solutions remain grounded in practical manufacturing considerations while leveraging the latest advances in machine learning technology (Figure 4.11).

The implementation of AI-driven systems must carefully balance the desire for process optimization with the need to maintain regulatory compliance and process control. Modern approaches utilize hybrid systems that combine artificial intelligence with traditional process control strategies, enabling automated optimization while maintaining human oversight of critical decisions. The development of these hybrid systems requires careful consideration of regulatory requirements, validation

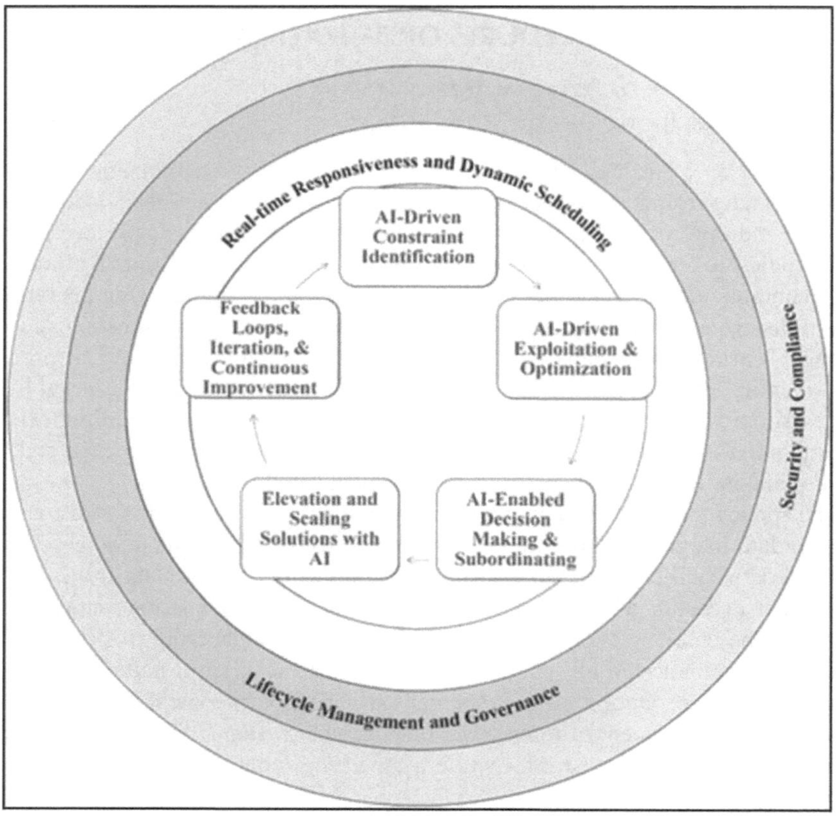

FIGURE 4.11 AI-driven SOP optimization framework.

protocols, and change control procedures to ensure that AI-driven improvements can be effectively implemented within GMP manufacturing environments.

The impact of artificial intelligence on operator training and process execution represents another critical aspect of modern manufacturing operations. AI-powered training systems can provide personalized instruction and real-time guidance while maintaining consistent documentation of operator qualifications and competency assessments. The development of these training systems requires careful consideration of learning objectives, assessment criteria, and documentation requirements to ensure effective knowledge transfer while maintaining regulatory compliance.

4.7.2 Machine Learning Applications in Process Parameter Optimization

The application of machine learning algorithms to process parameter optimization represents a fundamental advancement in biotechnology manufacturing, enabling more sophisticated approaches to process understanding and control. Modern facilities utilize various machine learning techniques, including supervised learning, unsupervised learning, and reinforcement learning, to analyze historical process data and identify optimal operating conditions. These approaches extend beyond traditional statistical analysis to incorporate complex nonlinear relationships and temporal dependencies that impact process performance and product quality (Table 4.12).

The development of predictive models for process optimization requires careful consideration of data quality, model validation, and implementation strategies. Modern approaches utilize sophisticated validation techniques that ensure model reliability while maintaining regulatory compliance. The implementation of these models must consider not only immediate process optimization objectives but also long-term manufacturing strategies and regulatory requirements. This necessitates careful documentation of model development, validation procedures, and ongoing performance monitoring to ensure continued effectiveness.

The integration of real-time optimization capabilities represents a significant advancement in process control strategies. Modern systems utilize online machine learning algorithms that can adapt to changing process conditions while maintaining consistent product quality. The development of these adaptive systems requires

TABLE 4.12
Machine Learning Applications in Bioprocess Optimization

Application Area	ML Technique	Primary Benefits	Implementation Challenges
Parameter optimization	Deep learning	Improved productivity	Model validation
Process monitoring	Anomaly detection	Early deviation detection	False positive control
Quality prediction	Regression models	Reduced testing time	Data quality requirements
Pattern recognition	Neural networks	Enhanced understanding	Model interpretability
Risk assessment	Decision trees	Proactive control	Regulatory acceptance

sophisticated approaches to model updating and validation, ensuring that optimization recommendations remain within established operating ranges while maintaining regulatory compliance. This becomes particularly critical in continuous manufacturing operations, where process conditions may evolve significantly over extended production periods.

Machine learning algorithms increasingly incorporate uncertainty quantification and risk assessment capabilities, enabling more informed decision-making in process optimization. These approaches utilize probabilistic models that provide not only optimization recommendations but also confidence intervals and risk assessments associated with different operating conditions. The development of these risk-aware systems requires careful consideration of regulatory requirements, validation protocols, and implementation strategies to ensure effective use in GMP manufacturing environments.

4.7.3 Natural Language Processing for SOP Documentation and Analysis

Natural language processing (NLP) has revolutionized the way biotechnology facilities manage and analyze standard operating procedures, introducing sophisticated capabilities for document analysis, version control, and compliance verification. Modern NLP systems can process extensive collections of SOPs, extracting key information, identifying potential inconsistencies, and ensuring alignment with regulatory requirements. The implementation of these systems requires careful consideration of document structure, terminology standardization, and validation requirements to ensure effective integration within existing quality management systems. The development of specialized NLP algorithms for biotechnology applications has enabled more sophisticated approaches to document analysis and management.

The application of NLP technologies in SOP management extends beyond simple text analysis to encompass complex semantic understanding and relationship mapping. Modern systems utilize advanced language models that can interpret technical terminology, recognize procedural dependencies, and identify potential gaps or conflicts in documentation. This sophisticated analysis enables more effective maintenance of SOP libraries while ensuring consistent terminology and procedural alignment across different documents. The development of these capabilities requires careful training of language models using domain-specific terminology and document structures typical of biotechnology manufacturing.

Document version control and change management represent critical applications of NLP technology in SOP management. Modern systems can automatically track modifications, analyze the impact of changes across related documents, and ensure consistent implementation of updates throughout the documentation system. The sophistication of these systems extends to the analysis of change implications, helping to identify potential impacts on related procedures and quality systems. This capability becomes particularly valuable in complex manufacturing environments where multiple SOPs may be interconnected through shared processes or requirements.

The integration of NLP capabilities with quality management systems has enabled more sophisticated approaches to compliance verification and audit preparation. Modern systems can automatically analyze SOPs against regulatory requirements, identifying potential compliance gaps and suggesting necessary updates. This automated analysis significantly reduces the manual effort required for compliance verification while ensuring more consistent and thorough review of documentation. The development of these compliance verification capabilities requires careful consideration of regulatory requirements across different jurisdictions and product types.

Translation and localization of SOPs represent another critical application of NLP technology in global manufacturing operations. Modern systems can assist in the accurate translation of technical documents while maintaining consistent terminology and procedural clarity across different languages. The implementation of these translation capabilities requires sophisticated approaches to technical terminology management and context preservation to ensure accurate communication of critical procedures across global manufacturing networks. This becomes particularly important in organizations operating multiple facilities across different regions with varying language requirements.

4.7.4 Computer Vision Applications in Process Monitoring and Control

The integration of computer vision technologies within biotechnology manufacturing represents a significant advancement in process monitoring and control capabilities. Modern facilities utilize sophisticated image analysis systems to monitor various aspects of manufacturing operations, from cell culture characteristics to equipment status and operator actions. The implementation of these systems requires careful consideration of imaging requirements, analysis algorithms, and validation protocols to ensure reliable operation in GMP manufacturing environments. The development of specialized computer vision applications for biotechnology has enabled more sophisticated approaches to process monitoring and quality control.

Real-time monitoring of cell culture characteristics represents a fundamental application of computer vision technology in upstream processing. Modern systems utilize advanced imaging techniques and analysis algorithms to assess cell morphology, density, and viability without compromising sterility or process integrity. The development of these monitoring capabilities requires sophisticated approaches to image acquisition, analysis, and data integration to provide meaningful insights into culture health and productivity. This becomes particularly critical in continuous manufacturing operations where rapid detection of process deviations is essential for maintaining product quality.

The application of computer vision in equipment monitoring and maintenance has enabled more proactive approaches to system management and contamination control. Modern facilities utilize automated imaging systems to monitor critical equipment components, detect potential issues before they impact process performance, and verify proper execution of cleaning and sterilization procedures. The implementation of these monitoring systems requires careful consideration of camera placement, lighting conditions, and analysis algorithms to ensure reliable detection of potential issues while minimizing false alarms.

Operator action verification represents another critical application of computer vision technology in GMP manufacturing environments. Modern systems can automatically monitor and document operator actions, ensuring compliance with established procedures while maintaining complete records of process execution. The development of these monitoring capabilities requires careful consideration of privacy requirements, validation protocols, and documentation standards to ensure effective implementation while maintaining operator acceptance and regulatory compliance.

4.7.5 AI-Driven Quality Control and Process Validation

The implementation of artificial intelligence in quality control and process validation represents a fundamental shift in how biotechnology facilities ensure product quality and process consistency. Modern approaches utilize sophisticated AI algorithms to analyze multiple quality parameters simultaneously, predict potential deviations before they impact product quality, and optimize sampling strategies based on historical process performance. The development of these capabilities requires careful consideration of regulatory requirements, validation protocols, and implementation strategies to ensure effective integration within existing quality systems.

Quality parameter prediction represents a critical application of AI technology in biotechnology manufacturing. Modern systems utilize advanced machine learning algorithms to predict final product quality based on in-process measurements and historical performance data. This predictive capability enables more proactive approaches to process control and quality assurance, potentially reducing the need for extensive end-product testing while maintaining product quality assurance. The implementation of these predictive systems requires careful validation to ensure reliable operation across different process conditions and product types.

Automated batch record review represents another significant application of AI technology in quality control operations. Modern systems can analyze complex batch records, identify potential compliance issues, and verify complete documentation of critical process parameters and controls. The sophistication of these systems extends to the analysis of trends and patterns across multiple batches, enabling more effective process optimization and quality improvement initiatives. The development of these capabilities requires careful consideration of regulatory requirements and validation protocols to ensure reliable operation in GMP environments.

Risk assessment and mitigation strategies have been significantly enhanced through the application of AI technologies in quality control operations. Modern systems utilize sophisticated algorithms to analyze historical process data, identify potential risk factors, and develop targeted mitigation strategies based on observed patterns and relationships. The implementation of these risk assessment capabilities requires careful consideration of regulatory requirements, validation protocols, and documentation standards to ensure effective integration within existing quality systems.

Statistical process control has evolved significantly through the integration of AI capabilities, enabling more sophisticated approaches to process monitoring and control. Modern systems utilize advanced algorithms to analyze process trends, detect subtle deviations, and predict potential quality issues before they impact product specifications. The development of these capabilities requires careful consideration of statistical validity, regulatory requirements, and implementation strategies to ensure effective use in GMP manufacturing environments.

4.8 AUTOMATING AND OPTIMIZING SOPS USING AI TECHNOLOGIES

4.8.1 Integration of Artificial Intelligence in Modern Biotechnology Operations

The integration of artificial intelligence technologies into standard operating procedures represents a paradigm shift in biotechnology manufacturing, fundamentally transforming how processes are designed, monitored, and optimized. Modern biotechnology facilities increasingly leverage sophisticated AI systems that encompass machine learning algorithms, natural language processing capabilities, and advanced pattern recognition tools to enhance operational efficiency and maintain product quality. This technological revolution extends beyond simple automation to include intelligent systems capable of learning from historical data, predicting process deviations, and recommending optimal operating conditions in real time. The implementation of AI-driven systems requires careful consideration of existing regulatory frameworks while pushing the boundaries of traditional manufacturing approaches to achieve unprecedented levels of process control and optimization.

The foundation of successful AI integration lies in the development of robust data infrastructure capable of capturing, processing, and analyzing vast amounts of process information in real time. Modern facilities must implement sophisticated data management systems that integrate information from multiple sources, including process sensors, analytical instruments, and quality control systems. This comprehensive data architecture enables AI systems to develop accurate models of process behavior while identifying subtle relationships between operating parameters and product quality attributes. The establishment of reliable data pipelines becomes particularly critical when implementing AI systems in GMP environments, where data integrity and traceability requirements must be rigorously maintained.

Contemporary AI implementations in biotechnology manufacturing increasingly utilize hybrid approaches that combine multiple artificial intelligence technologies to achieve optimal results. These systems may incorporate supervised learning algorithms for process prediction, unsupervised learning for pattern detection, and reinforcement learning for process optimization. The selection and implementation of appropriate AI technologies must consider not only immediate process requirements but also long-term scalability and validation objectives. The development of hybrid AI systems requires careful attention to algorithm selection, model validation, and system maintenance to ensure reliable operation across different process conditions and scale ranges (Figure 4.12).

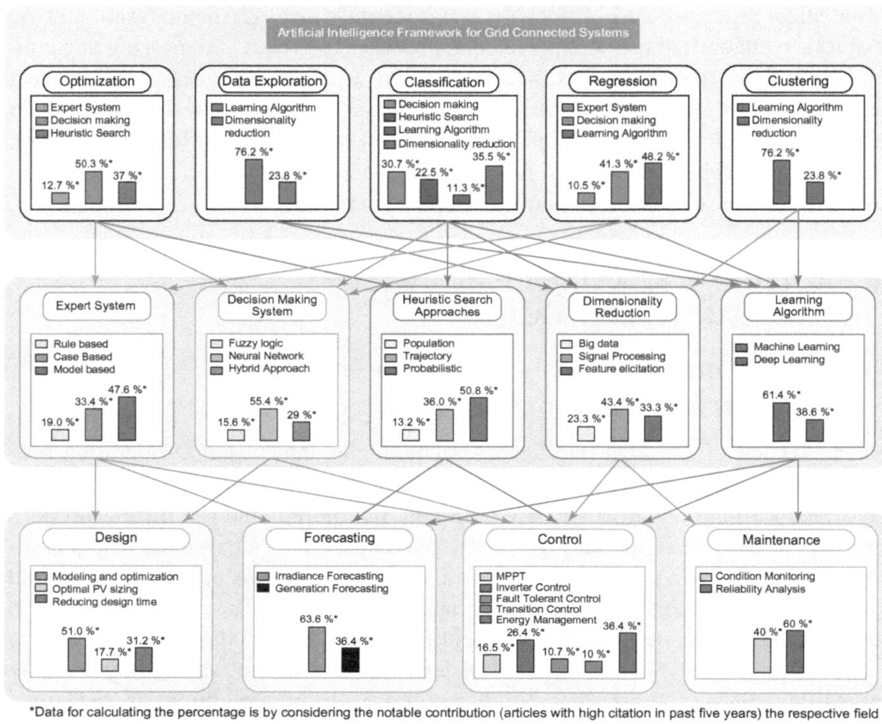

FIGURE 4.12 Integrated AI framework for SOP optimization.

The implementation of AI-driven systems fundamentally changes the role of human operators in biotechnology manufacturing, requiring careful consideration of training requirements and operational protocols. Modern facilities must develop comprehensive training programs that enable personnel to effectively interact with AI systems while maintaining critical thinking skills and process understanding. The evolution of operator roles extends beyond basic system operation to include data interpretation, model validation, and system optimization activities. This transformation requires careful attention to change management processes and organizational development strategies to ensure successful adoption of AI technologies.

The validation of AI systems presents unique challenges in regulated manufacturing environments, requiring novel approaches to system qualification and performance verification. Traditional validation approaches must be adapted to address the dynamic nature of AI systems while maintaining compliance with regulatory requirements. The development of validation strategies must consider not only immediate system performance but also long-term reliability and maintenance requirements. This validation process becomes particularly critical when implementing AI systems in GMP manufacturing environments, where system performance must be consistently demonstrated and documented.

TABLE 4.13
Machine Learning Applications in Biotech Manufacturing

Application Area	ML Algorithm Type	Primary Benefits	Implementation Challenges
Process prediction	Deep neural networks	Early deviation detection	Model validation
Pattern recognition	Convolutional networks	Quality pattern identification	Data preprocessing
Optimization	Reinforcement learning	Real-time parameter adjustment	System integration
Anomaly detection	Autoencoders	Early fault detection	Alert management
Process control	Hybrid systems	Adaptive control strategies	Regulatory compliance

4.8.2 Machine Learning Applications in Process Optimization

The application of machine learning algorithms in biotechnology process optimization represents a significant advancement in manufacturing efficiency and product quality control. Contemporary implementations utilize sophisticated neural networks and deep learning architectures capable of processing complex process data to identify optimal operating conditions and predict potential deviations. These systems analyze historical process data alongside real-time measurements to develop accurate models of process behavior and recommend appropriate control actions. The implementation of machine learning systems requires careful consideration of model architecture, training procedures, and validation requirements to ensure reliable operation in regulated manufacturing environments (Table 4.13).

The development of effective machine learning models requires sophisticated approaches to data preprocessing and feature selection, ensuring that input data accurately represents critical process parameters and quality attributes. Modern implementations utilize advanced data cleaning algorithms and feature extraction techniques to identify relevant process information while minimizing the impact of noise and measurement uncertainty. The selection of appropriate preprocessing strategies must consider not only immediate model performance requirements but also long-term stability and maintainability objectives. This preparation phase becomes particularly critical when implementing machine learning systems in regulated manufacturing environments, where data quality and integrity must be rigorously maintained.

4.9 CONCLUSION

The comprehensive examination of standard operating procedures (SOPs) in biotechnology manufacturing presented in this chapter underscores the critical importance of systematic approaches to biological product production. The evolution of SOPs from simple procedural documents to sophisticated, integrated systems reflects the increasing complexity of modern biotechnology manufacturing and the industry's commitment to ensuring product quality and patient safety. Through detailed analysis of upstream processing, downstream processing, quality control measures, and final fill-finish operations, we have demonstrated the interconnected nature of various manufacturing elements and the crucial role of standardized procedures in maintaining process consistency and product quality.

The implementation of effective SOPs in upstream processing has proven fundamental to successful biological product manufacturing, with careful attention required for cell line development, media optimization, and bioreactor operation. The sophisticated approaches to process monitoring and control discussed in this chapter highlight the industry's movement toward real-time process verification and continuous improvement methodologies. The integration of advanced analytical technologies and automated control systems has enabled unprecedented levels of process understanding and control, facilitating the consistent production of high-quality biological products.

Downstream processing operations have similarly benefited from the implementation of robust SOPs, with particular emphasis on purification strategy optimization and product quality maintenance. The careful integration of various unit operations within a comprehensive process control framework has enabled efficient production of therapeutic proteins while ensuring consistent product quality. The evolution of purification technologies and analytical methods has facilitated more efficient and effective downstream processing operations, though challenges remain in optimizing these processes for increasingly complex biological products.

Quality control systems have emerged as a critical component of modern biotechnology manufacturing, with SOPs playing a crucial role in ensuring consistent product quality throughout the manufacturing process. The implementation of sophisticated analytical methods and comprehensive monitoring systems has enabled more effective quality control, while the integration of risk-based approaches has enhanced the efficiency of quality assurance programs. The movement toward real-time release testing and continuous process verification represents a significant advancement in quality control methodology, though careful attention must be maintained to ensure regulatory compliance and product safety.

The final fill-finish operations discussed in this chapter demonstrate the critical importance of maintaining product quality through the final stages of manufacturing. The implementation of advanced aseptic processing technologies and automated filling systems has enhanced operational efficiency while maintaining strict control over product sterility and quality. The integration of sophisticated environmental monitoring systems and comprehensive contamination control strategies has further enhanced the reliability of fill-finish operations, though continued vigilance is required to ensure consistent product quality.

The examination of case studies involving therapeutic proteins, monoclonal antibodies, and human vaccines has provided practical insights into the application of SOPs in various manufacturing contexts. These examples have demonstrated the flexibility required in SOP implementation while maintaining strict adherence to quality and regulatory requirements. The diverse range of products and processes examined highlights the adaptability of modern manufacturing systems and the crucial role of well-designed SOPs in ensuring consistent product quality.

The integration of artificial intelligence and machine learning technologies in SOP optimization represents a significant advancement in manufacturing process control and improvement. These emerging technologies offer unprecedented capabilities for process understanding and optimization, though careful attention must be maintained to ensure appropriate validation and regulatory compliance. The continued evolution of these technologies promises to further enhance manufacturing efficiency and product quality while potentially reducing operational costs and development timelines.

5 Introduction to Quality Systems

5.1 INTRODUCTION

Quality systems in contemporary organizations represent far more than mere procedural frameworks or regulatory compliance mechanisms; they embody a comprehensive philosophical approach to organizational excellence that permeates every aspect of operations, decision-making, and strategic planning. The evolution of quality systems from their inception as simple inspection protocols to their current status as integrated, intelligence-driven frameworks reflects the profound transformation in how organizations conceptualize and implement quality management. This transformation encompasses not only technological advancements and methodological innovations but also fundamental shifts in organizational culture and operational paradigms. The modern quality system serves as a dynamic, adaptive framework that enables organizations to maintain consistency while fostering innovation, ensure compliance while promoting flexibility, and achieve efficiency while pursuing continuous improvement (Figure 5.1).

The historical development of quality systems provides crucial context for understanding their current form and future trajectory. Beginning with the industrial revolution's emphasis on standardization and inspection, quality management has undergone several revolutionary transformations. The statistical quality control movement of the early 20th century introduced scientific rigor to quality management, establishing mathematical foundations that continue to influence modern approaches. The post-World War II quality revolution, particularly in Japan, introduced concepts such as total quality management, emphasizing the importance of organization-wide commitment to quality principles. These historical developments have culminated in the current era of integrated quality systems, where traditional methodologies merge with cutting-edge technologies to create unprecedented capabilities for quality assurance and improvement.

The theoretical underpinnings of modern quality systems draw from a diverse array of disciplines, including systems theory, organizational psychology, operations research, and information science. This multidisciplinary foundation enables quality systems to address complex organizational challenges through various perspectives and methodological approaches. Systems thinking, in particular, has profoundly influenced quality system development, emphasizing the interconnectedness of organizational elements and the importance of considering both direct and indirect effects of quality-related decisions. The integration of psychological principles helps organizations understand and influence human factors in quality management, while

DOI: 10.1201/9781003624592-5

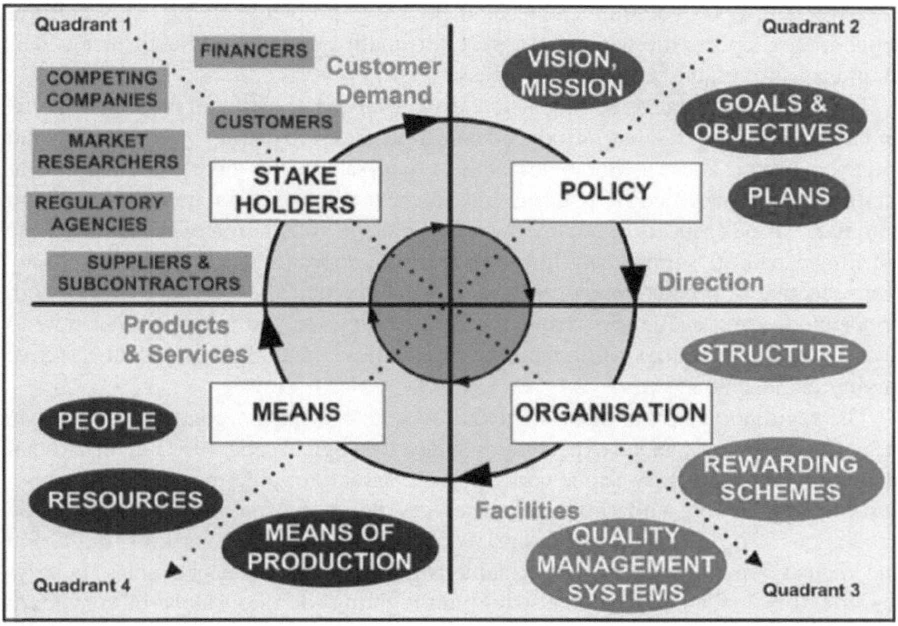

FIGURE 5.1 Quality system framework.

operations research provides analytical tools for optimizing system performance. This theoretical diversity creates a robust framework capable of addressing the multifaceted challenges of modern quality management.

The implementation of quality systems in contemporary organizations requires careful consideration of numerous contextual factors, including organizational structure, culture, technological infrastructure, and market environment. Successful quality systems must balance standardization with flexibility, control with innovation, and efficiency with effectiveness. This balancing act becomes particularly crucial in global organizations operating across different cultural, regulatory, and technological environments. The challenge of maintaining consistent quality standards while accommodating local variations has led to the development of sophisticated adaptation mechanisms within quality systems. These mechanisms enable organizations to maintain core quality principles while adjusting implementation approaches to suit specific operational contexts.

The relationship between quality systems and organizational culture represents a critical area of consideration in modern quality management. Effective quality systems both shape and are shaped by organizational culture, creating a dynamic interaction that influences system effectiveness and sustainability. This cultural dimension of quality systems extends beyond formal procedures and policies to encompass shared values, beliefs, and behaviors that support quality objectives. Organizations must cultivate a quality-oriented culture that encourages continuous improvement, open communication, and proactive problem-solving. This cultural foundation supports the technical aspects of quality systems while ensuring their long-term sustainability and effectiveness.

The technological revolution, particularly the advent of artificial intelligence (AI) and advanced analytics, has introduced new dimensions to quality system design and implementation. These technologies enable organizations to process vast amounts of quality-related data, identify patterns and trends, predict potential quality issues, and automate various aspects of quality management. The integration of AI and machine learning capabilities transforms how organizations approach quality control, moving from reactive to predictive approaches. This technological evolution presents both opportunities and challenges, requiring organizations to carefully consider how to leverage new capabilities while maintaining the fundamental principles of effective quality management.

The regulatory environment surrounding quality systems continues to evolve, reflecting changing societal expectations and technological capabilities. International standards and regulatory frameworks provide essential guidelines for quality system implementation while allowing necessary flexibility for organization-specific adaptations. The relationship between various standards, such as ISO 9001:2000 and industry-specific requirements, demonstrates the universal principles underlying effective quality management while highlighting the importance of contextual adaptation. Organizations must navigate this complex regulatory landscape while maintaining system effectiveness and efficiency, ensuring compliance without sacrificing innovation and adaptability.

The financial implications of quality systems represent another crucial consideration in modern organizations. While the implementation and maintenance of comprehensive quality systems require significant investment, the long-term benefits in terms of reduced errors, improved efficiency, and enhanced customer satisfaction typically justify these investments. Organizations must carefully evaluate the cost-benefit relationships associated with various quality system elements, considering both direct and indirect costs and benefits. This economic dimension of quality systems influences decision-making regarding system scope, technological investments, and resource allocation.

The human dimension of quality systems remains paramount despite technological advances. The role of leadership in establishing and maintaining effective quality systems cannot be overstated. Leaders must demonstrate visible commitment to quality principles, allocate necessary resources, and create an environment that supports continuous improvement. Additionally, the involvement and engagement of all organizational members in quality initiatives remains crucial for system success. Training, communication, and motivation strategies play essential roles in ensuring effective human participation in quality system implementation and maintenance.

Looking toward the future, quality systems continue to evolve in response to changing organizational needs and technological capabilities. The integration of artificial intelligence, blockchain technology, and other emerging technologies promises to further transform how organizations approach quality management. These technological advances, combined with evolving regulatory requirements and changing customer expectations, will require organizations to maintain adaptable, responsive quality systems capable of addressing new challenges while maintaining fundamental quality principles.

5.2 MAIN ELEMENTS OF A QUALITY SYSTEM

5.2.1 Foundational Framework of Quality Systems

The foundational framework of quality systems encompasses a complex network of interrelated elements that work in harmony to ensure organizational excellence and continuous improvement. At its core, this framework represents the systematic integration of processes, procedures, and methodologies that collectively define how an organization approaches quality management. The complexity of modern quality systems necessitates a deep understanding of how various elements interact and influence each other, creating a dynamic ecosystem that responds to both internal and external stimuli. This foundational framework serves as the backbone of quality management, providing structure while maintaining the flexibility necessary for adaptation to changing circumstances and requirements.

In examining the foundational framework, it becomes evident that successful quality systems are built upon multiple layers of interconnected components, each serving specific functions while contributing to the overall system effectiveness. The primary layer consists of organizational commitment and leadership support, without which no quality system can truly flourish. This commitment manifests through resource allocation, strategic planning, and the establishment of clear quality objectives that align with organizational goals. The secondary layer encompasses the structural elements of the quality system, including documentation systems, process controls, and measurement methodologies. These structural elements provide the tangible mechanisms through which quality principles are implemented and maintained throughout the organization.

The relationship between various framework elements creates a synergistic effect that enhances overall system performance. This synergy is particularly evident in how different components support and reinforce each other, creating a robust system capable of addressing diverse quality challenges. For instance, the documentation system supports process control by providing clear guidelines and procedures, while measurement methodologies provide feedback that enables continuous improvement of both documentation and processes. This interconnected nature of framework elements ensures that the quality system functions as a cohesive unit rather than a collection of isolated component (Figure 5.2).

5.2.2 Documentation and Record Management Systems

Documentation within quality systems serves as the cornerstone of organizational knowledge management and operational consistency. A comprehensive documentation system encompasses various levels of documentation, from high-level policy statements to detailed work instructions and records. The structure of documentation

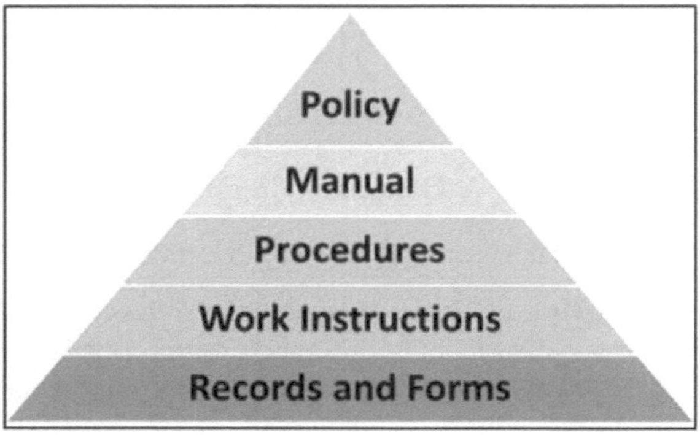

FIGURE 5.2 Quality system framework hierarchy.

follows a hierarchical pattern that ensures proper flow of information and maintains consistency across different organizational levels. This hierarchical approach to documentation management enables organizations to maintain control over their processes while providing necessary flexibility for operational requirements.

The importance of effective documentation management extends beyond mere recordkeeping to encompass knowledge preservation and transfer within the organization. Well-maintained documentation systems serve as repositories of organizational knowledge, capturing best practices, lessons learned, and critical process information. This knowledge preservation function becomes particularly crucial in modern organizations where workforce mobility and technological change create challenges for maintaining operational continuity. The documentation system must therefore balance the need for comprehensive information capture with accessibility and usability considerations.

Modern documentation systems leverage advanced technologies to enhance their effectiveness and efficiency. Digital document management systems, equipped with sophisticated search and retrieval capabilities, enable organizations to maintain vast repositories of quality-related information while ensuring quick access when needed. These technological solutions also facilitate version control, document distribution, and collaborative editing, ensuring that quality system documentation remains current and accessible to all relevant stakeholders. The integration of artificial intelligence and machine learning capabilities further enhances documentation management by enabling automated categorization, intelligent search, and predictive document routing.

5.2.3 Process Control and Monitoring Mechanisms

Process control represents a fundamental aspect of quality systems, encompassing the methods and tools used to maintain consistent process performance and output quality. Effective process control requires a deep understanding of process parameters, their interactions, and their impact on final output quality. Organizations must establish robust monitoring mechanisms that enable real-time tracking of process performance and early detection of potential quality issues. These monitoring mechanisms must balance the need for comprehensive data collection with practical considerations regarding resource allocation and operational efficiency.

The evolution of process control methodologies reflects the increasing complexity of modern manufacturing and service operations. Traditional statistical process control methods have been enhanced with advanced monitoring technologies that enable more sophisticated approaches to process management. These technologies include sensor networks, real-time data analytics, and automated control systems that provide unprecedented capabilities for process monitoring and control. The integration of these technologies enables organizations to maintain tighter control over their processes while reducing the resource requirements associated with manual monitoring and intervention.

The relationship between process control and quality improvement represents a critical aspect of quality system effectiveness. Effective process control not only maintains current quality levels but also provides the foundation for continuous improvement initiatives. By generating detailed process performance data, control systems enable organizations to identify improvement opportunities and evaluate the effectiveness of improvement actions. This data-driven approach to process improvement ensures that quality enhancement efforts are based on objective evidence rather than subjective assessments.

5.2.4 Resource Management and Allocation

Resource management within quality systems encompasses the identification, allocation, and optimization of various organizational resources required for effective quality management. These resources include human capital, technological infrastructure, financial resources, and physical assets. The effective management of these resources requires careful consideration of both current operational needs and future quality improvement objectives. Organizations must develop comprehensive resource management strategies that ensure optimal resource utilization while maintaining the flexibility to address changing quality requirements.

Human resource management plays a particularly crucial role in quality system effectiveness. The knowledge, skills, and commitment of organizational members directly impact the success of quality initiatives. Organizations must invest in training and development programs that enhance quality-related competencies while fostering a culture of continuous improvement. These programs should address both technical skills related to specific quality tools and methodologies and soft skills such as problem-solving, communication, and teamwork. The development of a quality-oriented workforce requires sustained commitment to employee development and engagement.

5.2.5 Performance Measurement and Analysis

Performance measurement systems provide the quantitative foundation for quality system evaluation and improvement. These systems must capture relevant data

TABLE 5.1
Quality System Performance Measurement Dimensions

Dimension	Key Metrics	Measurement Frequency
Process quality	Defect rates, process capability indices	Continuous/daily
Customer satisfaction	Customer feedback scores, complaint rates	Monthly/quarterly
System effectiveness	Audit results, improvement project success rates	Quarterly/annual
Resource efficiency	Resource utilization rates, training effectiveness	Monthly/quarterly

across various organizational dimensions, enabling comprehensive assessment of quality system effectiveness. The selection of appropriate performance metrics requires careful consideration of organizational objectives, stakeholder requirements, and practical measurement constraints. Organizations must develop balanced measurement frameworks that address both leading and lagging indicators of quality performance (Table 5.1).

5.2.6 Continuous Improvement Framework

The continuous improvement framework represents the dynamic aspect of quality systems, providing mechanisms for systematic enhancement of system effectiveness. This framework encompasses various methodologies and tools for identifying improvement opportunities, implementing changes, and evaluating their effectiveness. The success of continuous improvement initiatives depends on the integration of various quality system elements, including performance measurement, process control, and resource management. Organizations must establish structured approaches to improvement that balance the need for systematic methodology with the flexibility to address diverse improvement opportunities.

Effective continuous improvement requires active engagement from all organizational levels and functions. Leadership must demonstrate visible commitment to improvement initiatives while providing necessary resources and removing organizational barriers. Middle management plays a crucial role in translating strategic improvement objectives into operational actions, while frontline employees contribute valuable insights and suggestions based on their direct process experience. This multilevel engagement ensures that improvement efforts address both strategic priorities and operational realities.

5.3 ESSENTIALS OF A QUALITY SYSTEM

5.3.1 Fundamental Framework Components

The foundational elements of quality systems encompass a complex network of interconnected components that collectively ensure organizational excellence and continuous improvement. At its core, a quality system's fundamental framework extends far beyond simple documentation and procedures, incorporating sophisticated mechanisms for control, evaluation, and enhancement of organizational processes. The framework begins with the establishment of clear quality objectives that align with organizational strategy and stakeholder requirements. These objectives serve as guideposts for system development and implementation, ensuring that all quality-related activities contribute meaningfully to organizational success. The integration of these objectives throughout the organizational hierarchy requires careful consideration of various factors, including operational capabilities, resource constraints, and market demands.

The structural integrity of quality systems relies heavily on the establishment of robust governance mechanisms that define roles, responsibilities, and accountability measures throughout the organization. This governance structure must balance the need for centralized control with the flexibility required for effective operation across different organizational units and geographical locations. The implementation of such governance mechanisms requires careful consideration of organizational culture, existing power structures, and communication patterns. Successful quality systems incorporate governance models that promote transparency, facilitate decision-making, and ensure consistent application of quality principles across all organizational levels.

Quality system frameworks must also address the critical aspect of resource allocation and management. This encompasses not only financial and material resources but also human capital, technological infrastructure, and knowledge management systems. The effective deployment of resources within quality systems requires sophisticated planning and control mechanisms that ensure optimal utilization while maintaining system flexibility. Organizations must develop comprehensive resource management strategies that account for both current operational needs and future development requirements, ensuring sustainable system operation and evolution.

5.3.2 Documentation and Control Mechanisms

Documentation within quality systems serves as more than a mere recordkeeping function; it represents a crucial mechanism for knowledge preservation, communication, and continuous improvement. The documentation hierarchy in modern quality systems encompasses various levels of information, from high-level policy documents to detailed operational procedures and work instructions. This hierarchical structure ensures that quality-related information is appropriately organized and accessible to relevant stakeholders while maintaining necessary control over document creation, modification, and distribution. The evolution of documentation systems has been significantly influenced by technological advancements, particularly in digital document management and control systems.

The control mechanisms embedded within quality systems extend beyond traditional document control to encompass various operational aspects, including process control, change management, and corrective action systems. These mechanisms must be designed to maintain system stability while facilitating necessary adaptations to changing requirements and conditions. The implementation of effective control mechanisms requires careful consideration of various factors, including organizational structure, technological capabilities, and human factors. Modern quality systems incorporate sophisticated control mechanisms that leverage advanced technologies while maintaining user-friendly interfaces and clear operational guidelines.

The integration of documentation and control mechanisms within broader organizational systems represents a critical challenge in quality system implementation. Organizations must ensure that quality-related documentation and control processes align effectively with other business processes and systems, avoiding unnecessary duplication or conflicts. This integration requires careful consideration of various interfaces and dependencies, ensuring smooth information flow and consistent application of quality principles across different organizational functions and levels.

5.3.3 Performance Measurement and Analysis

The measurement and analysis of quality system performance represent critical functions that enable organizations to evaluate system effectiveness and identify improvement opportunities. Performance measurement frameworks must encompass various dimensions of quality, including product quality, process efficiency, customer satisfaction, and system compliance. The development of comprehensive measurement systems requires careful selection of appropriate metrics and indicators that provide meaningful insights into system performance while avoiding excessive complexity or data overload.

Modern quality systems incorporate sophisticated analytical tools and methodologies that enable organizations to derive meaningful insights from performance data. These analytical capabilities extend beyond traditional statistical analysis to include advanced techniques such as predictive analytics and pattern recognition. The integration of these analytical capabilities within quality systems requires careful consideration of data quality, analytical tool selection, and result interpretation methodologies. Organizations must ensure that analytical processes provide actionable insights while maintaining accuracy and reliability (Table 5.2).

5.3.4 Continuous Improvement Mechanisms

Continuous improvement represents a fundamental principle of modern quality systems, requiring systematic approaches to identifying and implementing enhancement

TABLE 5.2
Quality System Performance Metrics Framework

Metric Category	Key Indicators	Measurement Frequency	Analysis Methods
Process quality	Defect rates, cycle time	Continuous	Statistical process control
System effectiveness	Compliance rates, audit results	Quarterly	Trend analysis
Customer satisfaction	Feedback scores, complaint rates	Monthly	Comparative analysis

opportunities. The establishment of effective improvement mechanisms requires careful consideration of various factors, including organizational culture, resource availability, and technological capabilities. Organizations must develop comprehensive improvement frameworks that encourage innovation while maintaining system stability and effectiveness.

The implementation of continuous improvement mechanisms requires robust methodologies for problem identification, root cause analysis, and solution development. These methodologies must be integrated within broader quality system frameworks while maintaining sufficient flexibility to address diverse improvement opportunities. The success of improvement initiatives depends heavily on effective stakeholder engagement and clear communication of improvement objectives and outcomes.

5.3.5 Risk Management Integration

The integration of risk management principles within quality systems represents a critical aspect of modern quality management. Organizations must develop comprehensive approaches to identifying, assessing, and managing quality-related risks across various operational domains. The implementation of effective risk management strategies requires careful consideration of various factors, including regulatory requirements, stakeholder expectations, and operational constraints.

Risk management frameworks within quality systems must address both internal and external risk factors while maintaining appropriate balance between risk mitigation and operational efficiency. The development of such frameworks requires sophisticated approaches to risk assessment, prioritization, and control. Organizations must ensure that risk management processes align effectively with other quality system elements while maintaining necessary flexibility for addressing emerging risks and challenges.

5.3.6 Stakeholder Engagement and Communication

Effective stakeholder engagement represents a crucial element of successful quality systems, requiring comprehensive approaches to identification, communication, and management of stakeholder relationships. Organizations must develop sophisticated strategies for engaging various stakeholder groups, including employees, customers, suppliers, and regulatory authorities. The implementation of such strategies requires careful consideration of communication channels, feedback mechanisms, and engagement methodologies.

The establishment of effective communication frameworks within quality systems requires consideration of various factors, including organizational culture, technological capabilities, and stakeholder preferences. Organizations must ensure that communication processes facilitate clear understanding of quality requirements and expectations while maintaining appropriate levels of transparency and accessibility. The success of stakeholder engagement initiatives depends heavily on consistent application of communication strategies and regular evaluation of engagement effectiveness.

5.4 PRACTICAL IMPLEMENTATION OF A QUALITY SYSTEM

5.4.1 Foundational Implementation Strategies

The practical implementation of quality systems represents one of the most critical challenges facing modern organizations, requiring a carefully orchestrated approach that balances theoretical principles with operational realities. Implementation strategies must account for the complex interplay between organizational structure, human resources, technological capabilities, and business objectives. The foundation of successful quality system implementation lies in the development of a comprehensive understanding of organizational context, including both internal capabilities and external factors that might influence system effectiveness. This contextual analysis enables organizations to develop implementation approaches that align with their specific needs while maintaining consistency with established quality management principles.

The initial phase of quality system implementation demands extensive preparation and planning, encompassing everything from resource allocation to stakeholder engagement strategies. Organizations must conduct thorough assessments of their current quality management practices, identifying gaps between existing procedures and desired outcomes. This assessment phase requires active participation from all organizational levels, ensuring that the resulting implementation strategy reflects diverse perspectives and operational requirements. The development of detailed implementation timelines and resource allocation plans provides a structured framework for system deployment while maintaining necessary flexibility to address unforeseen challenges.

Stakeholder engagement represents a crucial element of successful quality system implementation, requiring careful attention to communication strategies and participation mechanisms. Organizations must develop comprehensive approaches to engaging various stakeholder groups, from top management to frontline employees, ensuring that all participants understand their roles and responsibilities within the new quality framework. This engagement process extends beyond simple information dissemination to include active participation in system design and implementation decisions, fostering a sense of ownership and commitment among stakeholders.

5.4.2 Documentation and Process Integration

Documentation plays a pivotal role in quality system implementation, serving as both a guide for system deployment and a record of organizational knowledge. The development of comprehensive documentation requires careful attention to detail while maintaining accessibility and usability for various stakeholder groups. Organizations must strike a delicate balance between providing sufficient detail for effective system operation and avoiding excessive complexity that might hinder implementation efforts. The documentation structure should reflect organizational hierarchy while facilitating information flow across different operational levels.

Process integration represents another critical aspect of quality system implementation, requiring careful attention to existing operational workflows and procedures. Organizations must develop detailed strategies for integrating quality system requirements with existing processes, ensuring minimal disruption to ongoing operations while achieving desired quality objectives. This integration process demands careful consideration of process interdependencies and potential impacts on operational efficiency. The development of clear integration protocols helps organizations maintain operational continuity while implementing new quality management approaches.

5.4.3 Training and Competency Development

The human dimension of quality system implementation cannot be overstated, with training and competency development serving as crucial elements of successful deployment. Organizations must develop comprehensive training programs that address both technical and cultural aspects of quality system implementation. These programs should provide participants with necessary knowledge and skills while fostering understanding of quality management principles and objectives. The development of role-specific training modules ensures that all organizational members receive appropriate preparation for their responsibilities within the quality system framework.

5.4.4 Monitoring and Evaluation Mechanisms

The establishment of effective monitoring and evaluation mechanisms represents a critical component of quality system implementation, enabling organizations to assess implementation progress and system effectiveness. These mechanisms must provide accurate and timely information about system performance while facilitating continuous improvement efforts. The development of comprehensive performance metrics requires careful consideration of both quantitative and qualitative factors, ensuring balanced assessment of system effectiveness (Table 5.3).

5.4.5 Continuous Improvement Integration

The integration of continuous improvement mechanisms within the implementation process ensures long-term system effectiveness and adaptability. Organizations must develop structured approaches to identifying improvement opportunities and implementing necessary changes. This improvement process requires active participation from all organizational levels, fostering a culture of continuous enhancement and innovation. The development of feedback mechanisms enables organizations to capture and utilize implementation experiences for system refinement and optimization.

TABLE 5.3
Critical Implementation Success Factors

Success Factor Category	Key Elements	Implementation Considerations
Leadership commitment	Strategic alignment, resource allocation	Visible support and participation
Stakeholder engagement	Communication, participation mechanisms	Inclusive decision-making processes
Resource management	Human, financial, technological resources	Efficient allocation and utilization
Process integration	Workflow analysis, integration protocols	Minimal operational disruption
Performance monitoring	Metrics development, evaluation systems	Balanced assessment approaches

5.4.6 Technology Integration and Automation

The role of technology in quality system implementation continues to evolve, with advanced solutions offering new opportunities for system optimization and automation. Organizations must carefully evaluate technological options, selecting solutions that align with their specific needs and capabilities. The integration of technological solutions requires careful attention to system compatibility and user requirements, ensuring effective adoption and utilization. The development of comprehensive technology integration strategies helps organizations maximize the benefits of advanced solutions while minimizing implementation challenges.

5.4.7 Risk Management and Mitigation

Risk management plays a crucial role in quality system implementation, requiring careful attention to potential challenges and mitigation strategies. Organizations must develop comprehensive risk assessment protocols, identifying potential implementation obstacles and developing appropriate response strategies. The integration of risk management principles throughout the implementation process helps organizations maintain system effectiveness while addressing potential challenges. The development of contingency plans ensures organizational readiness to address unexpected implementation issues while maintaining progress toward quality objectives.

5.5 STRUCTURE OF A QUALITY MANUAL

5.5.1 Foundational Elements of Quality Manual Architecture

The architecture of a quality manual represents the cornerstone of any robust quality management system, serving as the primary document that outlines an organization's quality philosophy, methodology, and implementation framework. The structural composition of a quality manual extends far beyond a mere collection of procedures and policies; it embodies the organization's commitment to quality excellence and provides a comprehensive road map for achieving and maintaining superior quality standards across all operational domains. In contemporary organizations, the quality manual serves as a living document that evolves continuously to reflect changing organizational needs, technological advancements, and regulatory requirements while maintaining its fundamental role as the authoritative source of quality system documentation.

The hierarchical organization of a quality manual reflects the complex interrelationships between various quality system components, creating a coherent framework that facilitates both understanding and implementation. At its core, the manual must establish clear lines of authority, responsibility, and accountability while simultaneously promoting flexibility and adaptability in response to changing circumstances. This delicate balance between structure and flexibility represents one of the most challenging aspects of quality manual development, requiring careful consideration of organizational culture, operational requirements, and strategic objectives.

The integration of risk management principles within the quality manual structure has become increasingly crucial in modern quality systems. Organizations must develop comprehensive approaches to identifying, assessing, and mitigating quality-related risks while maintaining operational efficiency and effectiveness. This risk-based thinking must permeate every aspect of the quality manual, from high-level policy statements to detailed procedural guidelines, ensuring that potential quality issues are addressed proactively rather than reactively.

5.5.2 Documentation Hierarchy and Control Systems

The documentation hierarchy within a quality manual establishes a systematic approach to information management and accessibility, ensuring that quality-related information flows effectively throughout the organization. This hierarchical structure typically encompasses multiple levels of documentation, ranging from high-level policy statements to detailed work instructions and records. The relationship between these various documentation levels must be clearly defined and maintained to prevent confusion and ensure consistent implementation of quality practices.

The control systems governing quality manual documentation play a vital role in maintaining document integrity and ensuring that only current, approved versions are in use throughout the organization. These control mechanisms must address various aspects of document management, including creation, review, approval, distribution, and revision. The implementation of electronic document management systems has revolutionized this process, enabling organizations to maintain better control over their quality documentation while improving accessibility and reducing administrative burden.

Modern quality manuals must incorporate provisions for electronic signatures and digital authentication methods, reflecting the increasing digitalization of quality management processes. These technological adaptations must be balanced with regulatory requirements and security considerations, ensuring that document

TABLE 5.4
Quality Manual Content Organization Framework

Section Category	Primary Focus	Key Elements	Implementation Considerations
Quality policy	Strategic direction	Vision, mission, objectives	Organizational alignment
System processes	Operational framework	Procedures, instructions	Process integration
Control measures	Performance monitoring	Metrics, standards	Effectiveness assessment

integrity and authenticity are maintained while leveraging the benefits of digital transformation.

5.5.3 Content Organization and Accessibility

The organization of content within a quality manual must facilitate both comprehension and implementation of quality requirements. This organizational structure should reflect logical progression from general principles to specific applications, enabling users to navigate the manual effectively and locate relevant information quickly. The manual's content must be presented in a clear, concise manner while maintaining sufficient detail to ensure proper understanding and implementation of quality requirements (Table 5.4).

The accessibility of quality manual content represents a critical consideration in modern organizations, particularly those operating across multiple locations or time zones. The manual must be readily available to all relevant personnel while maintaining appropriate security controls and access restrictions. Digital platforms and mobile accessibility have become increasingly important in ensuring that quality documentation remains accessible to authorized users regardless of location or time.

5.5.4 Integration with Management Systems

The integration of quality manual requirements with other management systems represents a crucial aspect of modern quality management. Organizations must ensure that quality system requirements align effectively with environmental management systems, occupational health and safety programs, and other specialized management frameworks. This integration helps prevent duplication of effort while ensuring comprehensive coverage of all relevant requirements.

The quality manual must establish clear interfaces between various management systems, defining how quality requirements interact with and support other organizational objectives. This integrated approach helps organizations optimize resource utilization while maintaining the effectiveness of individual management systems. The manual should provide guidance on resolving potential conflicts between different system requirements while ensuring that quality objectives remain paramount.

5.5.5 Review and Update Mechanisms

The establishment of effective review and update mechanisms ensures that the quality manual remains current and relevant to organizational needs. These mechanisms must address both scheduled periodic reviews and event-driven updates, ensuring that the manual reflects changes in regulatory requirements, technological capabilities, and organizational structure. The review process should incorporate input from various stakeholders, including management, quality professionals, and operational personnel.

The frequency and scope of quality manual reviews must be carefully balanced to maintain document currency without creating excessive administrative burden. Organizations must establish clear criteria for determining when updates are necessary and implement efficient processes for executing these updates while maintaining system integrity. The review process should include provisions for assessing the effectiveness of existing quality manual content and identifying opportunities for improvement.

5.5.6 Implementation Guidance and Training

The quality manual must provide comprehensive guidance for implementing quality system requirements effectively throughout the organization. This guidance should include detailed information on interpreting and applying quality requirements in various operational contexts, ensuring consistent implementation while allowing necessary flexibility for specific situations. The manual should also address training requirements and methodologies for ensuring that personnel understand and can effectively implement quality system requirements.

Organizations must develop effective training programs that enable personnel to understand and apply quality manual requirements effectively in their daily operations. These training programs should incorporate both theoretical knowledge and practical applications, ensuring that personnel can translate quality manual requirements into effective operational practices. The manual should provide guidance on developing and maintaining training programs while establishing criteria for assessing training effectiveness.

5.6 CORRELATION BETWEEN GMP REQUIREMENTS (WHO) AND ISO 9001:2000

5.6.1 Foundational Framework Integration

The correlation between Good Manufacturing Practice (GMP) requirements as established by the World Health Organization and ISO 9001:2000 standards represents a crucial intersection in modern quality management systems. This integration encompasses a complex web of complementary requirements, methodologies, and implementation strategies that organizations must navigate to ensure comprehensive quality assurance. The fundamental alignment between these two frameworks stems from their shared objective of ensuring consistent, high-quality outputs while maintaining systematic control over production processes. The WHO GMP requirements, primarily focused on pharmaceutical and healthcare-related manufacturing, provide specific guidelines for ensuring product safety and efficacy, while ISO 9001:2000 offers a broader quality management framework applicable across various industries. The synergy between these frameworks creates a robust foundation for quality system implementation that exceeds the capabilities of either framework in isolation.

Organizations operating in regulated industries, particularly those in pharmaceutical manufacturing and healthcare products, must understand the intricate relationships between GMP requirements and ISO 9001:2000 standards to develop effective quality systems. The integration of these frameworks requires careful consideration of various factors, including documentation requirements, process validation methodologies, risk management approaches, and continuous improvement mechanisms. The harmonization of these requirements enables organizations to establish comprehensive quality systems that satisfy both regulatory compliance needs and operational excellence objectives. This integration becomes particularly crucial in global operations where organizations must navigate multiple regulatory environments while maintaining consistent quality standards.

The evolution of quality management systems has led to increasing alignment between GMP requirements and ISO standards, reflecting a growing recognition of the benefits of harmonized approaches to quality management. This alignment manifests in various areas, including documentation hierarchies, risk assessment methodologies, and process validation approaches. Organizations can leverage this alignment to develop more efficient and effective quality systems that address both specific GMP requirements and broader quality management objectives. The convergence of these frameworks also facilitates the implementation of integrated management systems that can more effectively address the complex challenges of modern manufacturing environments (Figure 5.3).

5.6.2 Documentation System Harmonization

Documentation represents a critical area of overlap between GMP requirements and ISO 9001:2000 standards, with both frameworks emphasizing the importance of comprehensive documentation systems. The harmonization of documentation requirements between these frameworks enables organizations to develop integrated document management systems that satisfy both regulatory compliance needs and operational efficiency objectives. This integration encompasses various documentation levels, from high-level quality policies to detailed standard operating procedures and work instructions. The alignment of documentation requirements facilitates more effective knowledge management and process control while reducing redundancy and potential inconsistencies in quality system documentation.

The implementation of harmonized documentation systems requires careful consideration of various factors, including document hierarchy, control mechanisms, revision management, and accessibility requirements. Organizations must establish clear relationships between different types of documents while ensuring appropriate

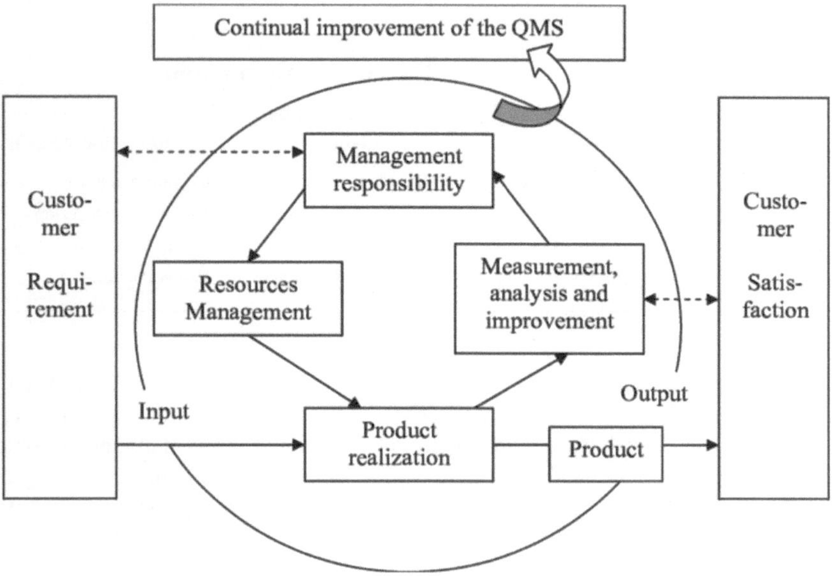

FIGURE 5.3 Integrated framework architecture.

levels of control and review for each document category. The integration of electronic document management systems has further transformed how organizations approach documentation harmonization, enabling more efficient document control and distribution while maintaining compliance with both GMP and ISO requirements. This technological evolution has introduced new possibilities for documentation system optimization while presenting unique challenges in areas such as electronic signatures and data integrity.

5.6.3 Process Validation and Control Integration

Process validation and control methodologies represent another crucial area of correlation between GMP requirements and ISO 9001:2000 standards. Both frameworks emphasize the importance of systematic approaches to process validation, monitoring, and control, although they may differ in specific requirements and terminology. The integration of validation approaches from both frameworks enables organizations to develop comprehensive validation strategies that satisfy regulatory requirements while promoting operational excellence. This integration becomes particularly important in industries where product quality directly impacts public health and safety, requiring robust validation approaches that combine GMP-specific requirements with ISO quality management principles.

The development of integrated validation strategies requires careful consideration of various factors, including validation scope, methodology selection, acceptance criteria establishment, and documentation requirements. Organizations must

TABLE 5.5
Integrated Quality Risk Management Framework Components

Component	GMP Focus	ISO 9001:2000 Alignment	Integration Benefits
Risk assessment	Product safety	Process quality	Comprehensive coverage
Risk control	Compliance	System effectiveness	Enhanced protection
Risk communication	Regulatory	Stakeholder	Complete transparency
Risk review	Safety impact	System performance	Holistic evaluation

balance the specific validation requirements of GMP guidelines with the broader process management principles of ISO 9001:2000, creating validation approaches that effectively address both regulatory compliance and quality management objectives. The implementation of risk-based validation approaches, incorporating elements from both frameworks, enables organizations to optimize resource allocation while maintaining appropriate levels of control over critical processes.

5.6.4 Quality Risk Management Integration

Quality risk management represents a fundamental aspect of both GMP requirements and ISO 9001:2000 standards, with both frameworks emphasizing the importance of systematic approaches to risk identification, assessment, and mitigation. The integration of risk management approaches from these frameworks enables organizations to develop comprehensive risk management strategies that address both product-specific safety concerns and broader quality management objectives. This integration becomes particularly crucial in regulated industries where product quality directly impacts public health and safety, requiring robust risk management approaches that combine specific GMP requirements with ISO quality management principles (Table 5.5).

5.6.5 Continuous Improvement Mechanism Integration

The integration of continuous improvement mechanisms represents a critical aspect of correlating GMP requirements with ISO 9001:2000 standards. Both frameworks emphasize the importance of systematic approaches to quality improvement, although they may differ in specific methodologies and focus areas. The harmonization of improvement approaches enables organizations to develop comprehensive improvement strategies that address both regulatory compliance requirements and operational excellence objectives. This integration becomes particularly important in industries where continuous improvement directly impacts product quality and safety, requiring robust approaches that combine GMP-specific requirements with ISO quality management principles.

5.6.6 Management Review and Oversight Integration

Management review and oversight processes represent another crucial area of correlation between GMP requirements and ISO 9001:2000 standards. Both frameworks emphasize the importance of systematic management review processes, although they may differ in specific requirements and focus areas. The integration of management review approaches enables organizations to develop comprehensive oversight mechanisms that effectively address both regulatory compliance requirements and quality management objectives. This integration becomes particularly important in ensuring effective leadership engagement in quality system maintenance and improvement, requiring approaches that combine specific GMP requirements with ISO management principles.

5.7 ARTIFICIAL INTELLIGENCE IN QUALITY SYSTEMS

5.7.1 Fundamentals of AI Integration in Quality Management

The integration of artificial intelligence into quality systems represents a transformative advancement in how organizations approach quality management and control. This revolutionary merger of traditional quality methodologies with cutting-edge AI technologies has fundamentally altered the landscape of quality assurance and continuous improvement. The implementation of AI in quality systems extends far beyond simple automation, encompassing sophisticated predictive analytics, real-time monitoring capabilities, and adaptive learning systems that continuously evolve to meet changing quality requirements. This technological integration enables organizations to achieve unprecedented levels of quality control while simultaneously reducing costs and improving operational efficiency. The foundational aspects of AI integration in quality systems include the development of neural network architectures specifically designed for quality monitoring, the implementation of machine learning algorithms capable of detecting subtle patterns in quality metrics, and the creation of automated response systems that can initiate corrective actions without human intervention.

The transformation of traditional quality systems through AI integration has led to the emergence of what many industry experts term "Quality 4.0," representing the fourth major revolution in quality management practices. This new paradigm leverages the power of artificial intelligence to create self-learning quality systems that can anticipate potential quality issues before they manifest, automatically adjust process parameters to maintain optimal quality levels, and provide detailed insights into quality trends that might be invisible to human observers. The implementation of AI-driven quality systems requires a fundamental rethinking of traditional quality management approaches, moving from reactive to proactive methodologies that emphasize prevention over correction.

5.7.2 Machine Learning Applications in Quality Control

The application of machine learning algorithms in quality control processes has revolutionized how organizations approach quality assurance and maintenance.

Deep learning networks, particularly convolutional neural networks (CNNs), have demonstrated exceptional capability in visual inspection tasks, surpassing human accuracy in many cases. These systems can process thousands of products per minute, identifying defects and variations that might be imperceptible to the human eye. The implementation of machine learning in quality control extends beyond visual inspection to encompass predictive maintenance, process optimization, and root cause analysis. Advanced algorithms can analyze vast amounts of historical quality data to identify patterns and correlations that can help prevent future quality issues.

The success of machine learning applications in quality control depends heavily on the quality and quantity of training data available. Organizations must develop comprehensive data collection and management strategies to ensure their AI systems have access to reliable, representative datasets. This includes implementing sophisticated data preprocessing techniques, developing robust data validation methods, and establishing clear protocols for data governance and management. The challenge of maintaining data quality becomes particularly crucial in environments where product specifications and quality requirements frequently change, requiring continuous updates to the AI system's training data.

5.7.3 Real-Time Quality Monitoring and Predictive Analytics

Real-time quality monitoring systems powered by AI have transformed how organizations approach quality control and assurance. These systems utilize advanced sensors and Internet of Things (IoT) devices to collect continuous streams of quality-related data, which is then analyzed in real time by sophisticated AI algorithms. The ability to process and analyze quality data in real time enables organizations to detect and respond to quality issues as they develop, rather than discovering problems after they have already impacted product quality. This proactive approach to quality management significantly reduces waste, improves efficiency, and enhances overall product quality.

The predictive analytics capabilities of AI-driven quality systems represent a particularly valuable advancement in quality management. By analyzing historical quality data and current process parameters, these systems can predict potential quality issues before they occur, allowing organizations to take preventive action. This predictive capability extends to maintenance requirements, material usage optimization, and process parameter adjustments, creating a comprehensive approach to quality management that minimizes disruptions and maximizes efficiency.

5.7.4 Natural Language Processing in Documentation and Compliance

The application of natural language processing (NLP) technologies in quality management systems has revolutionized how organizations handle documentation, compliance reporting, and quality-related communication. NLP-powered systems can automatically analyze quality documentation, extract relevant information, and identify potential compliance issues or documentation gaps. This capability significantly reduces the time and effort required for documentation review and compliance monitoring while improving the accuracy and consistency of these processes.

The implementation of NLP in quality systems extends to the analysis of customer feedback, supplier communications, and internal quality reports. Advanced NLP algorithms can process vast amounts of unstructured text data to identify trends, potential quality issues, and opportunities for improvement. This capability enables organizations to maintain a more comprehensive understanding of their quality performance and stakeholder perspectives, facilitating more informed decision-making and more effective quality improvement initiatives.

5.7.5 AI-Driven Process Optimization

The optimization of quality-related processes through AI represents a significant advancement in how organizations approach continuous improvement. AI systems can analyze complex process parameters and their interactions to identify optimal operating conditions that maximize quality while minimizing resource consumption. These systems utilize sophisticated optimization algorithms that can consider multiple objectives simultaneously, balancing quality requirements with cost constraints, energy efficiency, and other operational considerations.

The implementation of AI-driven process optimization extends beyond traditional statistical process control methods to encompass dynamic optimization approaches that can adapt to changing conditions in real time. These systems can automatically adjust process parameters based on current quality metrics, environmental conditions, and production requirements, ensuring consistent quality even in highly variable operating environments. The ability to maintain optimal process conditions continuously without human intervention represents a significant advancement in quality management capabilities.

5.7.6 Collaborative AI Systems in Quality Management

The development of collaborative AI systems that can work alongside human quality professionals represents an important advancement in quality management practices. These systems combine the analytical power of artificial intelligence with human expertise and judgment, creating a synergistic approach to quality management that leverages the strengths of both human and machine intelligence. Collaborative AI systems can provide real-time decision support to quality professionals, offering data-driven insights and recommendations while allowing human operators to make final decisions based on their experience and contextual understanding.

The implementation of collaborative AI systems requires careful consideration of human factors and interface design to ensure effective human–machine interaction. Organizations must develop appropriate training programs to help quality professionals understand how to work effectively with AI systems, interpret AI-generated recommendations, and maintain appropriate oversight of automated processes. The success of collaborative AI systems depends heavily on achieving the right balance between automation and human control, ensuring that AI technology enhances rather than replaces human expertise in quality management.

5.8 LEVERAGING AI TO ENSURE CONTINUOUS COMPLIANCE AND IMPROVEMENT IN QUALITY SYSTEMS

5.8.1 Foundation of AI Integration in Quality Systems

The integration of artificial intelligence within quality management systems represents a revolutionary advancement in how organizations approach quality control, compliance monitoring, and continuous improvement initiatives. This transformation extends far beyond simple automation, encompassing sophisticated machine learning algorithms, natural language processing capabilities, and advanced predictive analytics that fundamentally alter the landscape of quality management. The incorporation of AI technologies enables organizations to shift from reactive quality management approaches to proactive, predictive methodologies that anticipate and prevent quality issues before they manifest. This paradigm shift has profound implications for organizational efficiency, resource allocation, and overall quality performance metrics.

The fundamental architecture of AI-driven quality systems builds upon traditional quality management frameworks while introducing dynamic, adaptive capabilities that respond to real-time data and changing operational conditions. This integration requires careful consideration of existing quality management processes, organizational culture, and technological infrastructure to ensure seamless implementation and maximum effectiveness. Organizations must develop comprehensive strategies that address not only the technical aspects of AI integration but also the human factors, cultural considerations, and change management requirements necessary for successful implementation. The transformation process involves careful evaluation of existing quality management processes, identification of areas where AI can provide maximum value, and development of implementation strategies that minimize disruption while maximizing benefits.

The systematic approach to AI integration in quality systems necessitates careful consideration of data quality, system architecture, and integration methodologies. Organizations must establish robust data collection and management processes that ensure the accuracy, completeness, and relevance of information feeding into AI systems. This includes developing comprehensive data governance frameworks, establishing data quality metrics, and implementing validation processes that maintain data integrity throughout the quality management life cycle. The architecture of AI-driven quality systems must support both current operational requirements and future scalability needs, allowing organizations to expand and enhance their capabilities as technology evolves and business needs change.

5.8.2 Predictive Analytics and Quality Forecasting

The implementation of predictive analytics within quality systems represents a significant advancement in organizations' ability to anticipate and prevent quality issues before they impact operations. This capability transforms traditional quality management approaches by introducing sophisticated forecasting models that analyze historical data, identify patterns, and predict potential quality issues with unprecedented accuracy. The predictive capabilities extend across multiple dimensions of quality management, including process control, maintenance scheduling, and resource allocation, enabling organizations to optimize their quality management efforts proactively.

Advanced machine learning algorithms analyze vast quantities of historical quality data, identifying subtle patterns and correlations that might escape human observation. These insights enable organizations to develop more effective quality control strategies, optimize resource allocation, and implement preventive measures before quality issues arise. The predictive capabilities extend beyond simple trend analysis, encompassing complex multivariate relationships and temporal patterns that influence quality outcomes. This sophisticated analysis enables organizations to develop more nuanced and effective quality management strategies that address both immediate concerns and long-term quality objectives.

5.8.3 Real-Time Monitoring and Adaptive Response Systems

The implementation of real-time monitoring capabilities through AI-driven systems represents a fundamental shift in how organizations approach quality control and compliance management. These systems continuously analyze operational data, comparing current performance against established quality parameters and identifying potential deviations before they become significant issues. The real-time monitoring capability enables immediate response to quality concerns, reducing the impact of quality issues and minimizing resource waste. Advanced pattern recognition algorithms identify subtle variations in process parameters that might indicate emerging quality issues, enabling proactive intervention before problems escalate.

The adaptive response capabilities of AI-driven quality systems enable automatic adjustments to process parameters based on real-time analysis of quality metrics. This dynamic adjustment capability ensures optimal process performance while maintaining quality standards across varying operational conditions. The system's ability to learn from historical data and adapt to changing conditions improves over time, leading to increasingly efficient and effective quality management processes. The integration of machine learning algorithms enables systems to identify complex relationships between process parameters and quality outcomes, facilitating more sophisticated control strategies that maintain optimal performance across diverse operating conditions.

5.8.4 Automated Documentation and Compliance Management

One of the most significant advantages of AI integration in quality systems lies in its ability to automate documentation and compliance management processes. Advanced natural language processing capabilities enable automated generation and validation of quality documentation, ensuring consistency and completeness while reducing the administrative burden on quality management personnel. The automation extends to compliance monitoring, where AI systems continuously verify adherence to regulatory requirements and internal quality standards, flagging potential compliance issues for immediate attention.

The automated documentation capabilities of AI-driven quality systems extend beyond simple recordkeeping, encompassing sophisticated analysis of documentation content, identification of potential gaps or inconsistencies, and automatic generation of compliance reports. This automation significantly reduces the time and resources required for documentation management while improving the accuracy and completeness of quality records. The system's ability to analyze documentation patterns enables identification of potential compliance risks and opportunities for process improvement, supporting both regulatory compliance and continuous improvement initiatives.

5.8.5 Continuous Learning and System Optimization

The implementation of continuous learning capabilities represents a crucial aspect of AI-driven quality systems, enabling ongoing optimization of quality management processes based on accumulated experience and emerging patterns. These systems analyze historical quality data, process parameters, and intervention outcomes to identify opportunities for improvement and optimize quality management strategies. The continuous learning capability ensures that quality systems become increasingly effective over time, adapting to changing conditions and emerging challenges while maintaining optimal performance.

The optimization process extends beyond simple parameter adjustment, encompassing sophisticated analysis of system behavior, identification of complex relationships between quality factors, and development of increasingly effective control strategies. This continuous improvement capability ensures that quality systems remain effective and efficient even as operational conditions change and new challenges emerge. The integration of machine learning algorithms enables systems to identify subtle patterns and relationships that might escape human observation, leading to more sophisticated and effective quality management strategies.

5.8.6 Integration with Legacy Systems and Future Scalability

The successful implementation of AI-driven quality systems requires careful consideration of integration requirements with existing quality management infrastructure and future scalability needs. Organizations must develop comprehensive integration strategies that ensure seamless communication between AI components and legacy systems while maintaining data integrity and system reliability. The integration process must address both technical and operational considerations, ensuring that AI capabilities enhance rather than disrupt existing quality management processes.

The scalability considerations extend beyond simple system capacity, encompassing the ability to adapt to changing business requirements, incorporate new technologies, and support expanding quality management needs. Organizations must develop flexible architectures that support both current operational requirements and future growth potential, ensuring that quality systems can evolve alongside organizational needs and technological capabilities.

5.9 CONCLUSION

The comprehensive exploration of quality systems presented in this chapter reveals the intricate interplay between theoretical foundations, practical implementations, and technological advancements in modern quality management. Through detailed examination of the main elements, essentials, and practical implementation strategies of quality systems, we have established a thorough understanding of how these systems function as integral components of organizational excellence. The correlation between GMP requirements and ISO 9001:2000 standards demonstrates the evolving nature of quality management frameworks and their adaptation to meet increasingly complex regulatory and operational demands.

The integration of artificial intelligence and advanced technologies in quality systems represents a transformative development in quality management practices. This technological evolution has fundamentally altered how organizations approach quality assurance, moving from reactive to predictive methodologies. The successful implementation of AI-driven quality systems has demonstrated the potential for enhanced efficiency, improved accuracy, and more sophisticated approaches to quality control and improvement. However, this integration also presents new challenges that organizations must carefully navigate, including data quality management, algorithm validation, and system integration considerations.

The examination of quality system structures and their practical implementation has revealed several critical insights for organizations seeking to establish or enhance their quality management frameworks. First, the importance of developing comprehensive, integrated approaches that address both regulatory requirements and operational objectives cannot be overstated. Second, the role of organizational culture in supporting quality initiatives emerges as a crucial factor in system effectiveness. Third, the need for flexible, adaptable quality systems that can respond to changing market conditions and technological developments while maintaining fundamental quality principles becomes increasingly apparent.

The relationship between quality systems and organizational performance demonstrates the strategic importance of effective quality management in modern business environments. Organizations that successfully implement comprehensive quality systems achieve not only improved operational efficiency and product quality but also enhanced customer satisfaction and competitive advantage. The systematic approach to quality management, supported by appropriate technological tools and methodologies, enables organizations to maintain consistent quality standards while pursuing continuous improvement objectives.

The future of quality systems appears increasingly oriented toward integration with emerging technologies and methodologies. The continued evolution of artificial intelligence, machine learning, and data analytics capabilities promises to further transform quality management practices. Organizations must prepare for this evolution by developing flexible, adaptable quality systems that can incorporate new technologies while maintaining fundamental quality principles. The importance of human factors in quality management remains paramount, despite technological advancement, emphasizing the need for balanced approaches that combine technological capabilities with human expertise and judgment.

The examination of quality manual structures and documentation requirements highlights the critical role of systematic documentation in quality system effectiveness. The development of comprehensive, well-organized quality documentation enables organizations to maintain consistency in operations while facilitating knowledge transfer and process improvement. The integration of electronic documentation systems and advanced data management tools has transformed how organizations approach quality system documentation, offering new possibilities for efficiency and effectiveness.

The correlation between various quality standards and requirements demonstrates the increasing convergence of quality management approaches across different industries and regulatory frameworks. This convergence enables organizations to develop more efficient, integrated quality systems that satisfy multiple regulatory requirements while maintaining operational effectiveness. The harmonization of quality standards facilitates international trade and collaboration while ensuring consistent quality levels across global supply chains.

The implementation of quality systems in modern organizations requires careful consideration of various factors, including resource allocation, technological infrastructure, and organizational capabilities. Successful implementation depends on strong leadership commitment, effective communication strategies, and comprehensive training programs. Organizations must develop implementation approaches that balance immediate operational needs with long-term quality objectives, ensuring sustainable quality improvement over time.

The role of continuous improvement in quality systems emerges as a fundamental principle that underlies all aspects of quality management. Organizations must establish systematic approaches to improvement that integrate both incremental and breakthrough improvement methodologies. The successful implementation of continuous improvement initiatives requires appropriate organizational structures, effective measurement systems, and robust feedback mechanisms.

As we look toward the future, the evolution of quality systems will likely continue to be shaped by technological advancement, changing regulatory requirements, and evolving market demands. Organizations must maintain flexible, adaptable quality systems capable of incorporating new technologies and methodologies while preserving fundamental quality principles. The successful integration of artificial intelligence and advanced analytics capabilities will become increasingly critical for maintaining competitive advantage in global markets.

The insights and methodologies presented in this chapter provide a foundation for understanding and implementing effective quality systems in modern organizations. By carefully considering the various elements, requirements, and implementation strategies discussed, organizations can develop comprehensive quality systems that support both operational excellence and regulatory compliance objectives. The continued evolution of quality management practices, supported by technological advancement and improved understanding of organizational dynamics, promises to further enhance the effectiveness of quality systems in supporting organizational success.

6 Principles and Practice of GMP

Personnel and Premises

6.1 INTRODUCTION

The pharmaceutical manufacturing landscape has evolved significantly over the past decades, driven by technological advancements, regulatory requirements, and the increasing complexity of pharmaceutical products. At the heart of this evolution lies the fundamental principles of Good Manufacturing Practice (GMP), which establish the critical framework for ensuring product quality, safety, and efficacy. The interconnection between personnel management and premises design represents one of the most crucial aspects of GMP implementation, forming an intricate web of requirements that must be carefully balanced and maintained.

The modern pharmaceutical manufacturing environment demands a sophisticated approach to personnel management that extends far beyond traditional human resource practices. This comprehensive approach encompasses not only the basic elements of recruitment and training but also delves deeply into the psychological aspects of quality culture development, continuous education programs, and the implementation of effective communication systems. The human element in pharmaceutical manufacturing requires careful consideration of individual competencies, team dynamics, and organizational structure, all of which must align perfectly with the technical requirements of pharmaceutical production.

Simultaneously, the design and maintenance of pharmaceutical manufacturing premises play a pivotal role in ensuring product quality and regulatory compliance. The physical infrastructure must support both the technical requirements of pharmaceutical production and the human factors that influence manufacturing excellence. This includes considerations of material flow, personnel movement, contamination control, and the integration of advanced manufacturing technologies. The relationship between personnel and premises is particularly evident in areas such as cleanroom operations, where human behavior and facility design must work in perfect harmony to maintain the required environmental conditions (Figure 6.1).

The evolution of pharmaceutical manufacturing has been marked by significant technological advancements, particularly in the realm of artificial intelligence (AI) and automation. These technologies have transformed both personnel management and premises optimization, introducing new possibilities for efficiency and compliance. The integration of AI-driven systems has revolutionized traditional approaches to training, documentation, and facility monitoring while simultaneously creating new challenges in terms of workforce adaptation and system validation.

DOI: 10.1201/9781003624592-6

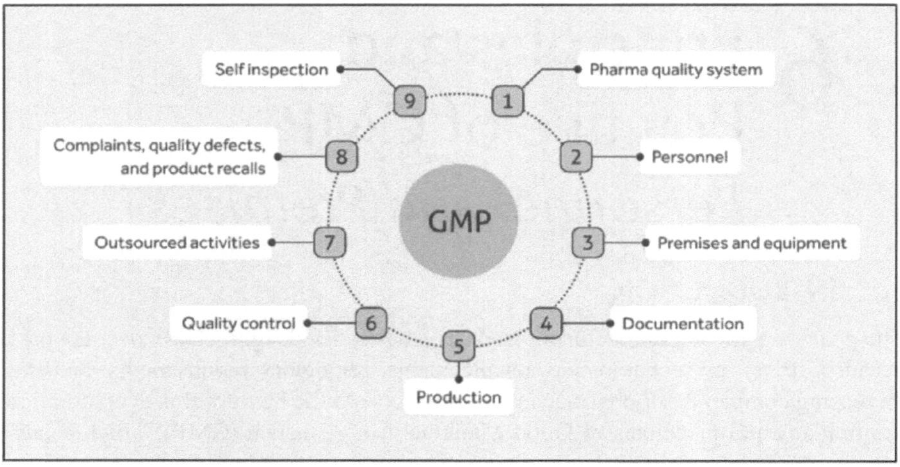

FIGURE 6.1 Integrated framework of GMP personnel and premises management description

The regulatory landscape surrounding pharmaceutical manufacturing continues to evolve, with increasing emphasis on risk-based approaches to quality management. This evolution necessitates a dynamic understanding of how personnel and premises requirements intersect with quality risk management principles. The implementation of effective risk management strategies requires a deep understanding of both human factors and technical capabilities, as well as the ability to adapt to changing regulatory requirements while maintaining consistent product quality.

The successful implementation of GMP principles in pharmaceutical manufacturing requires a holistic understanding of how human resources and physical infrastructure work together to ensure product quality and regulatory compliance. This chapter explores the intricate relationships between these elements, providing manufacturing professionals with comprehensive guidance on developing and maintaining effective systems for personnel and premises management. Through detailed examination of organizational structures, qualification requirements, facility specifications, and emerging technologies, this chapter serves as an essential resource for those responsible for implementing and maintaining GMP compliance in pharmaceutical manufacturing facilities.

The integration of personnel management and premises design represents a critical aspect of pharmaceutical manufacturing excellence. This chapter provides a detailed examination of the principles, practices, and technologies that support effective GMP implementation, with particular emphasis on the human factors and technical requirements that must be carefully balanced to ensure consistent product quality. Through careful consideration of these elements, manufacturing organizations can develop robust systems that support both regulatory compliance and operational excellence.

6.2 PRINCIPLES OF HUMAN RESOURCE MANAGEMENT

6.2.1 Foundational Elements of GMP Personnel Management

The implementation of effective human resource management within Good Manufacturing Practice (GMP) environments requires a sophisticated understanding of both regulatory requirements and organizational psychology. Personnel management in pharmaceutical manufacturing transcends traditional human resource practices, incorporating elements of quality assurance, risk management, and continuous improvement. The foundation of GMP personnel management rests upon the principle that human capital represents the most critical factor in maintaining product quality and regulatory compliance. This understanding necessitates a comprehensive approach to personnel development that begins with recruitment and extends throughout the entire employment life cycle.

The development of robust personnel management systems requires careful consideration of organizational culture, technical competency requirements, and regulatory expectations. Manufacturing organizations must establish frameworks that support both individual growth and collective excellence while simultaneously maintaining strict compliance with GMP requirements. This complex balance demands careful attention to documentation systems, training programs, and performance evaluation mechanisms that align with both regulatory requirements and organizational objectives.

The integration of quality management principles into personnel development programs represents a critical aspect of GMP compliance. Organizations must establish systems that not only ensure technical competency but also foster a deep understanding of quality principles among all personnel. This integration requires careful consideration of training methodologies, assessment techniques, and continuous improvement mechanisms that support both individual development and organizational excellence. The establishment of quality-focused training programs must address not only technical skills but also the broader aspects of GMP compliance, including documentation practices, contamination control, and error prevention strategies.

Personnel qualification systems within GMP environments must incorporate both theoretical knowledge and practical capabilities. The development of comprehensive qualification programs requires careful consideration of job-specific requirements, regulatory expectations, and industry best practices. Organizations must establish clear criteria for initial qualification, ongoing assessment, and periodic requalification that ensure consistent compliance with GMP requirements. This systematic approach to personnel qualification supports both regulatory compliance and operational excellence while simultaneously providing clear pathways for professional development.

The implementation of effective communication systems represents another critical aspect of GMP personnel management. Organizations must establish mechanisms that support clear, timely, and accurate communication of quality-related information throughout the manufacturing operation. This includes both vertical communication between management levels and horizontal communication across functional areas. The development of robust communication systems requires careful consideration of organizational structure, reporting relationships, and information flow patterns that support effective decision-making and problem resolution processes.

6.2.2 Quality Culture Development and Maintenance

The establishment and maintenance of a robust quality culture represents one of the most challenging aspects of GMP personnel management. Quality culture development requires a comprehensive understanding of organizational behavior, change management principles, and regulatory requirements. Organizations must create environments that naturally foster quality consciousness and compliance awareness among all personnel, from entry-level operators to senior management. This cultural development process requires sustained effort and commitment from all organizational levels, supported by appropriate systems and structures.

Quality culture development begins with clear articulation of organizational values and expectations regarding product quality and regulatory compliance. Manufacturing organizations must establish frameworks that support consistent application of quality principles throughout all operational areas. This includes development of appropriate policies, procedures, and work instructions that clearly communicate quality expectations while simultaneously providing practical guidance for daily operations. The integration of quality principles into routine operations requires careful balance between procedural requirements and practical considerations.

The maintenance of quality culture requires ongoing attention to both individual and collective behaviors that influence product quality. Organizations must establish mechanisms for monitoring and reinforcing desired behaviors while simultaneously addressing practices that could compromise product quality. This includes development of appropriate recognition systems that reinforce positive contributions to quality culture as well as corrective action mechanisms that address behavioral patterns that could impact product quality or regulatory compliance.

Leadership engagement represents a critical factor in quality culture development and maintenance. Senior management must demonstrate visible commitment to quality principles through both words and actions, establishing clear expectations regarding quality performance while simultaneously providing necessary resources for quality system maintenance. The development of effective leadership behaviors requires careful attention to both technical competency and interpersonal skills that support quality culture development.

6.2.3 Performance Management and Continuous Improvement

Performance management within GMP environments requires careful integration of quality metrics with traditional performance indicators. Organizations must establish systems that effectively evaluate both technical competency and compliance awareness while simultaneously supporting continuous improvement initiatives. The

TABLE 6.1
Key Performance Dimensions and Their Impact on Quality in GMP Environments

Performance Dimension	Key Indicators	Evaluation Frequency	Impact on Quality
Technical competency	Knowledge assessment results	Quarterly	Direct
GMP compliance	Deviation reports	Monthly	Critical
Documentation	Record accuracy	Weekly	High
Team effectiveness	Project outcomes	Semiannual	Indirect

development of comprehensive performance management systems requires careful consideration of regulatory requirements, organizational objectives, and individual development needs (Table 6.1).

The implementation of effective performance management systems requires establishment of clear performance standards that align with both regulatory requirements and organizational objectives. These standards must address not only technical competency requirements but also behavioral expectations regarding quality consciousness and regulatory compliance. The development of appropriate performance standards requires careful consideration of job-specific requirements, regulatory expectations, and industry best practices.

Continuous improvement initiatives within GMP environments must address both technical skills development and quality system enhancement. Organizations must establish mechanisms for identifying improvement opportunities, implementing appropriate changes, and evaluating the effectiveness of improvement initiatives. This systematic approach to continuous improvement requires careful integration of performance data, quality metrics, and regulatory requirements that support sustainable organizational development.

6.2.4 Knowledge Management and Transfer

Knowledge management within pharmaceutical manufacturing environments requires sophisticated systems for capturing, maintaining, and transferring critical information. Organizations must establish mechanisms that effectively preserve technical knowledge while simultaneously supporting development of new capabilities. The implementation of effective knowledge management systems requires careful consideration of both explicit and tacit knowledge transfer requirements.

The development of comprehensive knowledge management systems must address both technical and quality-related knowledge requirements. Organizations must establish mechanisms for capturing critical knowledge from experienced personnel while simultaneously supporting development of new capabilities among less experienced staff. This knowledge transfer process requires careful attention to documentation systems, training methodologies, and mentoring programs that support effective capability development.

6.2.5 Error Prevention and Human Factors Engineering

Human factors engineering in pharmaceutical manufacturing environments requires careful consideration of both physical and cognitive aspects of work performance. Organizations must establish systems that effectively support error prevention while simultaneously maintaining operational efficiency. The development of effective error prevention strategies requires careful integration of human factors principles with GMP requirements.

The implementation of error prevention strategies must address both systematic and random error potential. Organizations must establish mechanisms for identifying potential error sources, implementing appropriate controls, and monitoring system effectiveness. This comprehensive approach to error prevention requires careful consideration of human capabilities, limitations, and behavioral patterns that influence work performance.

6.2.6 Sustainability and Future Development

Sustainable development of personnel management systems requires careful attention to both current requirements and future needs. Organizations must establish frameworks that support ongoing system enhancement while maintaining consistent compliance with regulatory requirements. The development of sustainable personnel management systems requires careful integration of current best practices with emerging trends in pharmaceutical manufacturing.

6.3 DUTIES OF SENIOR MANAGEMENT

6.3.1 Strategic Quality Leadership

The role of senior management in pharmaceutical manufacturing extends far beyond traditional corporate leadership responsibilities, encompassing a complex web of quality-focused duties that directly impact product safety and regulatory compliance. In the context of Good Manufacturing Practice (GMP), senior management bears the ultimate responsibility for establishing and maintaining a robust quality management system that ensures consistent product quality while fostering a culture of continuous improvement. This responsibility manifests through various interconnected activities, including the development and implementation of quality policies, the allocation of resources for quality initiatives, and the establishment of clear communication channels throughout the organization.

Senior management must demonstrate an unwavering commitment to quality through both words and actions, setting the tone for the entire organization's approach to GMP compliance. This commitment involves regular engagement with quality-related activities, including participation in quality management reviews, oversight of corrective and preventive action (CAPA) systems, and direct involvement in critical quality decisions. The effectiveness of senior management's quality leadership can be observed through various indicators, including the organization's quality metrics, regulatory inspection outcomes, and the overall state of quality culture within the facility.

The establishment of clear quality objectives represents another crucial aspect of senior management's duties in GMP environments. These objectives must be specific, measurable, achievable, relevant, and time-bound, providing clear direction for all levels of the organization. Senior management must ensure that these objectives align with the organization's overall strategic goals while maintaining compliance with regulatory requirements and industry best practices. Regular review and adjustment of these objectives ensure their continued relevance and effectiveness in driving quality improvement.

6.3.2 Resource Management and Allocation

The effective allocation of resources stands as one of the most critical responsibilities of senior management in pharmaceutical manufacturing facilities. This responsibility encompasses the provision of adequate human resources, financial resources, and technical infrastructure necessary to maintain GMP compliance and ensure product quality. Senior management must demonstrate a thorough understanding of resource requirements across all aspects of pharmaceutical manufacturing, from routine operations to quality control activities and facility maintenance.

Resource allocation decisions must be based on a comprehensive assessment of organizational needs, regulatory requirements, and quality objectives. This assessment process requires senior management to maintain close communication with department heads, quality assurance personnel, and technical experts to understand resource requirements at all levels of the organization. The decision-making process must also consider long-term strategic objectives, ensuring that resource allocation supports both immediate operational needs and future growth opportunities.

Human resource allocation represents a particularly critical aspect of senior management's duties, requiring careful consideration of staffing levels, expertise requirements, and training needs. Senior management must ensure that all key positions are filled with qualified personnel and that adequate resources are available for ongoing training and development programs. This includes the provision of resources for GMP training, technical skill development, and leadership development programs that support the organization's quality objectives.

6.3.3 Quality Management System Oversight

Senior management bears direct responsibility for the establishment, maintenance, and continuous improvement of the organization's quality management system (QMS). This responsibility includes ensuring that the QMS is appropriately designed, effectively implemented, and regularly reviewed for effectiveness. The oversight of the QMS requires senior management to maintain active engagement with quality assurance activities, including regular review of quality metrics, audit findings, and regulatory inspection outcomes (Table 6.2).

TABLE 6.2
Senior Management Responsibilities and Key Performance Indicators for QMS Oversight

Management Responsibility	Key Performance Indicators	Review Frequency
Quality policy review	Policy effectiveness metrics	Annual
Resource allocation	Resource utilization rates	Quarterly
Quality objectives	Achievement of quality goals	Monthly
Management review	System effectiveness measures	Biannual
Compliance oversight	Regulatory compliance status	Continuous

The effectiveness of the QMS depends heavily on senior management's ability to foster a culture of quality throughout the organization. This involves promoting open communication about quality issues, encouraging employee participation in quality improvement initiatives, and demonstrating visible commitment to quality principles. Senior management must ensure that quality considerations are integrated into all aspects of organizational decision-making, from strategic planning to daily operations.

6.3.4 Regulatory Compliance and External Relations

Senior management plays a pivotal role in ensuring regulatory compliance and managing relationships with regulatory authorities. This responsibility includes maintaining awareness of regulatory requirements, ensuring adequate resources for compliance activities, and establishing effective systems for managing regulatory communications and inspections. Senior management must demonstrate a thorough understanding of the regulatory landscape and ensure that the organization maintains compliance with all applicable regulations and guidelines.

The management of regulatory relationships requires a proactive approach, including regular engagement with regulatory authorities, participation in industry associations, and monitoring of regulatory trends and developments. Senior management must ensure that the organization maintains effective systems for tracking and responding to regulatory changes, including the allocation of resources for implementing new regulatory requirements and updating existing systems as needed.

6.3.5 Change Management and Innovation

In the rapidly evolving pharmaceutical manufacturing landscape, senior management must effectively balance the need for innovation with the maintenance of GMP compliance. This requires careful oversight of change management processes, ensuring that proposed changes are thoroughly evaluated for their impact on product quality and regulatory compliance. Senior management must establish and maintain effective systems for managing change, including processes for change evaluation, implementation, and post-implementation review.

The promotion of innovation within GMP constraints requires senior management to foster a culture that encourages continuous improvement while maintaining strict adherence to quality and regulatory requirements. This involves supporting research and development initiatives, encouraging process improvement suggestions from employees, and ensuring that adequate resources are available for evaluating and implementing beneficial changes to manufacturing processes and systems.

6.3.6 Risk Management Integration

Senior management must ensure the effective integration of risk management principles throughout the organization's operations. This involves establishing comprehensive risk management systems that identify, assess, and mitigate potential risks to product quality and regulatory compliance. The implementation of risk management principles requires senior management to promote a risk-aware culture throughout the organization, ensuring that risk considerations are incorporated into all key decisions and processes.

Risk management integration extends beyond traditional quality risk management to encompass broader organizational risks, including operational risks, supply chain risks, and regulatory risks. Senior management must ensure that appropriate systems are in place for identifying and managing these risks, including regular risk assessments, the development and implementation of risk mitigation strategies, and the monitoring of risk management effectiveness.

6.4 ORGANIZATIONAL STRUCTURES

6.4.1 Hierarchical Framework in Pharmaceutical Manufacturing

The organizational structure within pharmaceutical manufacturing facilities represents a complex network of interconnected departments and roles, each contributing to the overall quality and compliance objectives of the organization. Traditional hierarchical structures have evolved to accommodate modern manufacturing requirements while maintaining clear lines of authority and responsibility. These structures must balance the need for efficient decision-making with the imperative of maintaining robust quality control systems. The implementation of effective organizational structures requires careful consideration of departmental relationships, reporting lines, and communication channels, all of which must align with GMP requirements and regulatory expectations.

Modern pharmaceutical organizations have moved beyond simple linear hierarchies to embrace matrix-based structures that facilitate cross-functional collaboration while maintaining clear accountability. This evolution reflects the increasing complexity of pharmaceutical manufacturing processes and the need for specialized expertise across multiple disciplines. Quality assurance departments, for instance, must maintain independence while simultaneously integrating with production, engineering, and research departments to ensure comprehensive quality oversight. The success of these organizational structures depends heavily on the clear definition of roles and responsibilities, coupled with effective communication systems that support rapid decision-making and problem resolution.

The integration of quality systems within organizational structures presents unique challenges that must be addressed through careful design and implementation of management systems. Quality units must maintain their independence while fostering collaborative relationships with other departments, ensuring that quality considerations remain paramount in all decision-making processes. This delicate balance requires sophisticated organizational designs that support both autonomy and integration, allowing quality personnel to exercise their authority effectively while maintaining productive working relationships across the organization.

6.4.2 Departmental Integration and Communication Flow

The effectiveness of pharmaceutical manufacturing organizations depends heavily on the seamless integration of various departments and functions. Production departments must work in close coordination with quality control, maintenance, and logistics teams to ensure efficient operations while maintaining compliance with GMP requirements. This integration requires sophisticated communication systems and well-defined procedures for cross-functional collaboration. The establishment of clear communication channels and decision-making protocols helps prevent delays and misunderstandings while ensuring that quality remains the primary focus of all operations.

Regular interdepartmental meetings and quality review sessions serve as crucial mechanisms for maintaining alignment across the organization. These forums provide opportunities for different departments to share information, address concerns, and collaborate on solutions to complex manufacturing challenges. The documentation of these interactions forms an essential part of the quality system, providing evidence of systematic communication and collaborative problem-solving approaches.

6.4.3 Quality Management Integration

The integration of quality management systems within the organizational structure represents a critical aspect of GMP compliance. Quality units must maintain appropriate independence while establishing effective working relationships with other departments. This independence ensures objective decision-making in matters relating to product quality and regulatory compliance. The organizational structure must support this independence while facilitating necessary collaboration between quality units and other operational departments (Table 6.3).

6.4.4 Document Management and Communication Systems

The organizational structure must support effective document management and communication systems that ensure the accurate and timely flow of information throughout the facility. These systems form the backbone of GMP compliance,

TABLE 6.3
Departmental Functions and Integration Points within the Quality Management System

Department	Primary Functions	Quality Integration Points	Communication Channels
Quality assurance	Quality system oversight	All departments	Formal and informal
Production	Manufacturing operations	Quality control, engineering	Standard operating procedures
Engineering	Technical support	Validation, maintenance	Technical reports
Research and development	Product development	Technology transfer	Project documentation

providing evidence of proper control and oversight of all manufacturing operations. The design of document management systems must consider the needs of various departments while maintaining compliance with regulatory requirements and internal quality standards.

Modern pharmaceutical organizations increasingly rely on electronic document management systems that facilitate rapid access to information while maintaining appropriate security and control measures. These systems must be validated to ensure reliability and compliance with regulatory requirements. The organizational structure must support the effective implementation and maintenance of these systems, including clear designation of responsibilities for system administration and document control.

6.4.5 Training and Development Integration

The organizational structure must support comprehensive training and development programs that ensure personnel maintain appropriate qualifications for their assigned responsibilities. These programs require careful coordination between human resources, quality assurance, and operational departments to ensure effective delivery and documentation of training activities. The structure must facilitate the identification of training needs, development of appropriate training materials, and evaluation of training effectiveness.

Training programs must address both technical skills and GMP requirements, ensuring that personnel understand both the practical aspects of their roles and the regulatory requirements that govern their activities. The organizational structure must support the delivery of these programs while maintaining appropriate documentation of all training activities. This includes systems for tracking training completion, evaluating competency, and identifying needs for refresher training.

6.4.6 Change Management and Continuous Improvement

The organizational structure must support effective change management processes that ensure controlled implementation of modifications to facilities, equipment, processes, and documentation. These systems require careful coordination between various departments to ensure that changes are properly evaluated, approved, and implemented. The structure must facilitate communication and collaboration throughout the change management process while maintaining appropriate controls and documentation.

Continuous improvement initiatives require organizational structures that support innovation while maintaining compliance with GMP requirements. These structures must facilitate the identification of improvement opportunities, evaluation of proposed changes, and implementation of approved modifications. The integration of continuous improvement programs within the organizational structure helps ensure sustained enhancement of manufacturing operations while maintaining appropriate quality controls.

6.5 QUALIFICATION AND PROFILES REQUIREMENT

6.5.1 Core Competency Framework for Pharmaceutical Personnel

The establishment of comprehensive qualification and profile requirements in pharmaceutical manufacturing represents a cornerstone of effective Good Manufacturing Practice implementation. The development of these requirements necessitates a thorough understanding of both technical and regulatory expectations, combined with practical considerations of operational efficiency and quality assurance. In modern pharmaceutical manufacturing facilities, the qualification framework must address not only the basic educational and experience requirements but also encompass the broader spectrum of competencies necessary for maintaining product quality and regulatory compliance.

The contemporary approach to qualification requirements in pharmaceutical manufacturing has evolved significantly from the traditional focus on academic credentials and technical expertise. Today's framework incorporates a multifaceted evaluation system that considers behavioral competencies, problem-solving abilities, and adaptability to technological advancement. This evolution reflects the increasing complexity of pharmaceutical manufacturing processes and the growing emphasis on quality culture development within manufacturing organizations. The integration of these various elements requires careful consideration of both individual capabilities and organizational needs, ensuring that personnel qualifications align with both current operational requirements and future developmental goals.

Quality assurance personnel in pharmaceutical manufacturing must possess a unique combination of technical knowledge and regulatory understanding. This requirement extends beyond basic familiarity with standard operating procedures and encompasses a deep comprehension of quality management systems, risk assessment methodologies, and regulatory compliance frameworks. The qualification requirements for these positions must reflect the critical nature of quality assurance responsibilities, including the ability to interpret and apply regulatory guidelines, conduct effective investigations, and implement corrective and preventive actions.

Manufacturing specialists require a distinct set of qualifications that combine theoretical knowledge with practical expertise in pharmaceutical production processes. These requirements must address not only the technical aspects of equipment operation and process control but also include understanding of contamination control principles, documentation requirements, and deviation management. The qualification framework for manufacturing personnel must additionally consider the increasing automation of production processes and the corresponding need for expertise in operating and maintaining sophisticated manufacturing equipment.

The validation of personnel qualifications represents another critical aspect of the qualification framework. This process must include both initial verification of credentials and ongoing assessment of competency maintenance. The validation process should incorporate practical evaluations, theoretical assessments, and regular review of performance metrics to ensure continued compliance with qualification requirements. This comprehensive approach to qualification validation helps ensure the maintenance of high quality standards throughout the manufacturing operation (Figure 6.2).

6.5.2 Technical Expertise Requirements

Technical expertise requirements in pharmaceutical manufacturing demand a sophisticated understanding of various scientific disciplines, including chemistry, microbiology, and engineering. The establishment of these requirements must consider the specific technical challenges associated with different manufacturing processes and product types. Personnel involved in technical operations must demonstrate

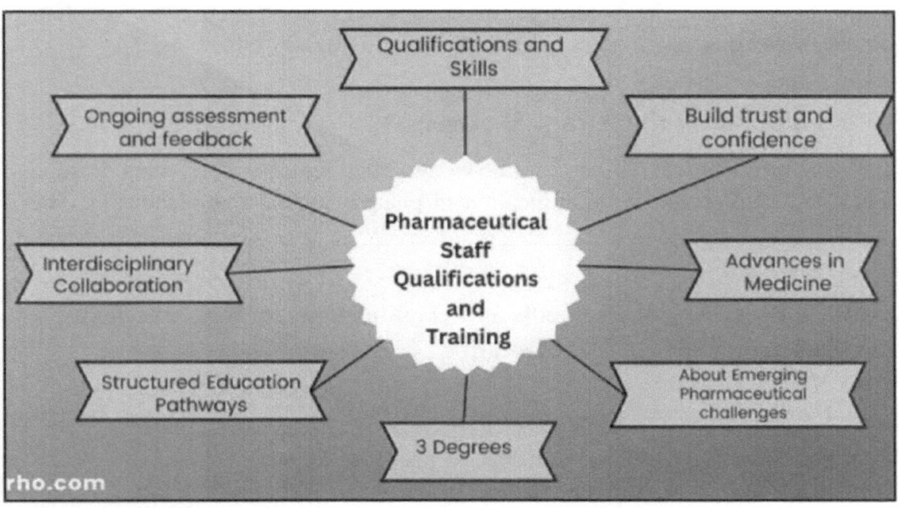

FIGURE 6.2 Comprehensive qualification framework for pharmaceutical personnel description.

proficiency in both theoretical concepts and practical applications, ensuring their ability to maintain product quality while optimizing manufacturing efficiency.

The development of technical expertise requirements must account for the increasing complexity of pharmaceutical manufacturing equipment and processes. Modern manufacturing facilities utilize sophisticated automation systems, analytical instruments, and process control technologies that require specialized technical knowledge. The qualification requirements for technical personnel must therefore include specific provisions for training and certification in the operation and maintenance of these advanced systems.

Process engineering expertise represents a critical component of technical qualification requirements. Personnel responsible for process development and optimization must possess advanced understanding of chemical engineering principles, process analytics, and quality by design concepts. These requirements should also address the ability to interpret and apply process analytical technology data, perform technical investigations, and implement process improvements while maintaining regulatory compliance.

Analytical expertise requirements form another crucial aspect of technical qualifications. Personnel involved in quality control and analytical testing must demonstrate proficiency in various analytical techniques, method validation procedures, and data interpretation. These requirements should encompass both routine analytical procedures and specialized testing methodologies, ensuring comprehensive coverage of all necessary technical capabilities.

Equipment qualification and maintenance expertise represents an essential component of technical requirements. Personnel responsible for equipment qualification and maintenance must possess detailed knowledge of equipment design principles, calibration procedures, and preventive maintenance protocols. These requirements should also address understanding of equipment qualification documentation, change control procedures, and impact assessment methodologies (Table 6.4).

6.5.3 Regulatory Knowledge Requirements

The establishment of regulatory knowledge requirements represents a fundamental aspect of personnel qualification in pharmaceutical manufacturing. These

TABLE 6.4
Technical Expertise Requirements and Qualifications for Key Functional Areas in Pharmaceutical Manufacturing

Technical Expertise Area	Required Qualifications	Experience Level	Specialized Training
Process engineering	Advanced degree	5+ years	PAT systems
Quality control	Scientific degree	3+ years	Analytical methods
Equipment maintenance	Technical degree	4+ years	Specific systems
Validation	Scientific/technical	4+ years	Validation types

requirements must address comprehensive understanding of various regulatory frameworks, including Good Manufacturing Practice guidelines, pharmaceutical quality systems, and international regulatory requirements. Personnel involved in regulatory compliance activities must demonstrate thorough knowledge of applicable regulations and the ability to interpret and apply regulatory guidance effectively.

Regulatory knowledge requirements must encompass understanding of quality management system requirements and their practical implementation. This includes knowledge of quality manual development, standard operating procedure creation, and quality system documentation maintenance. Personnel responsible for regulatory compliance must also demonstrate proficiency in conducting internal audits, managing regulatory inspections, and developing responses to regulatory observations.

The development of regulatory knowledge requirements must address understanding of change control procedures and their implementation. Personnel involved in change management must possess detailed knowledge of change impact assessment methodologies, documentation requirements, and regulatory notification procedures. These requirements should also encompass understanding of postapproval change management and its implications for regulatory compliance.

Qualification requirements for regulatory affairs personnel must include comprehensive understanding of registration requirements and regulatory submission procedures. This encompasses knowledge of common technical document preparation, regulatory agency communication protocols, and postapproval maintenance requirements. Personnel involved in regulatory affairs must also demonstrate proficiency in managing multiple regulatory submissions across different jurisdictions.

Regulatory knowledge requirements must address understanding of pharmacovigilance and product safety reporting requirements. Personnel involved in these activities must possess detailed knowledge of adverse event reporting procedures, safety signal detection methodologies, and regulatory reporting timelines. These requirements should also encompass understanding of risk management planning and implementation of risk minimization measures.

The validation of regulatory knowledge represents an ongoing requirement in pharmaceutical manufacturing. Personnel must maintain current understanding of evolving regulatory requirements and demonstrate ability to adapt compliance programs accordingly. This includes participation in continuing education programs, regulatory agency updates, and industry association activities focused on regulatory compliance maintenance.

6.5.4 Documentation and Recordkeeping Requirements

Documentation expertise requirements constitute a critical element of personnel qualification in pharmaceutical manufacturing. These requirements must address comprehensive understanding of documentation principles, including good documentation practices, electronic documentation systems, and data integrity requirements. Personnel involved in documentation activities must demonstrate proficiency in creating, reviewing, and maintaining various types of manufacturing documentation.

The establishment of documentation requirements must consider the increasing complexity of electronic documentation systems and data management tools. Personnel responsible for documentation management must possess detailed knowledge of electronic document management systems, data backup procedures, and electronic signature requirements. These requirements should also address understanding of data integrity principles and their practical implementation in documentation systems.

Recordkeeping requirements must encompass understanding of retention procedures and archival systems. Personnel involved in record management must demonstrate knowledge of record retention schedules, archival procedures, and retrieval systems. These requirements should also address understanding of documentation security measures and access control procedures for both paper-based and electronic records.

Documentation review requirements represent another critical aspect of qualification requirements. Personnel responsible for document review must possess detailed knowledge of review procedures, approval hierarchies, and documentation change control systems. These requirements should also encompass understanding of documentation life cycle management and periodic review procedures.

The development of documentation requirements must address understanding of batch record completion and review procedures. Personnel involved in batch documentation must demonstrate proficiency in accurate record completion, data verification procedures, and deviation documentation. These requirements should also encompass understanding of batch record review procedures and investigation documentation requirements.

6.5.5 Quality Culture and Behavioral Requirements

The establishment of quality culture requirements represents a fundamental aspect of personnel qualification in pharmaceutical manufacturing. These requirements must address behavioral competencies, communication skills, and commitment to quality principles. Personnel at all levels must demonstrate understanding of quality culture principles and their role in maintaining product quality and regulatory compliance.

Quality culture requirements must encompass understanding of error prevention methodologies and quality risk management principles. Personnel must demonstrate commitment to error prevention through consistent application of quality procedures, attention to detail, and proactive identification of potential quality issues. These requirements should also address understanding of human error reduction techniques and their practical implementation in daily operations.

The development of communication requirements represents another critical aspect of quality culture qualification. Personnel must demonstrate proficiency in both written and verbal communication, including ability to effectively communicate quality issues, participate in investigations, and contribute to quality improvement initiatives. These requirements should also encompass understanding of effective communication techniques for different organizational levels and situations.

Team collaboration requirements form an essential component of quality culture qualification. Personnel must demonstrate ability to work effectively in cross-functional teams, contribute to quality improvement projects, and participate in problem-solving activities. These requirements should also address understanding of team dynamics and effective collaboration techniques in quality-focused environments.

Leadership requirements within quality culture qualification must address ability to promote and maintain quality culture principles. Personnel in leadership positions must demonstrate commitment to quality through their actions, decisions, and communication style. These requirements should also encompass understanding of motivation techniques and their application in maintaining quality culture.

6.6 WORKPLACE AND JOB DESCRIPTIONS

6.6.1 Fundamentals of Workplace Design in GMP Environments

The design and organization of workplaces in pharmaceutical manufacturing facilities represent a critical intersection between operational efficiency and regulatory compliance. The modern pharmaceutical workplace must accommodate not only the technical requirements of manufacturing processes but also the human factors that influence product quality and safety. This comprehensive approach to workplace design begins with a thorough understanding of the manufacturing workflow, incorporating ergonomic principles while maintaining strict adherence to GMP requirements. The integration of these elements creates a working environment that supports both employee performance and product quality, establishing a foundation for consistent manufacturing excellence.

In pharmaceutical manufacturing environments, workplace design extends beyond basic ergonomic considerations to encompass a complex network of interrelated factors. These include material flow optimization, contamination control, personnel movement patterns, and emergency response capabilities. The successful integration of these elements requires careful consideration of both macro-level facility design and micro-level workstation configuration. This holistic approach ensures that each workspace not only supports efficient operation but also maintains the stringent quality standards required in pharmaceutical manufacturing.

The psychological aspects of workplace design play an increasingly important role in modern pharmaceutical manufacturing facilities. The creation of environments that promote focus, accuracy, and attention to detail while minimizing stress and fatigue has become a crucial consideration in workplace planning. This includes careful attention to lighting conditions, noise levels, temperature control, and visual organization of work areas. The implementation of visual management systems, including clear signage and color-coding schemes, supports consistent operation while reducing the potential for human error.

6.6.2 Job Description Development and Implementation

The development of comprehensive job descriptions in GMP environments requires a sophisticated understanding of both technical requirements and regulatory expectations. These documents serve as more than simple task lists; they represent critical quality system elements that define roles, responsibilities, and accountability within the pharmaceutical manufacturing organization. The creation of effective job descriptions begins with a detailed analysis of process requirements, incorporating both routine operations and exceptional circumstances that may require specialized knowledge or skills.

The implementation of job descriptions in pharmaceutical manufacturing environments requires careful consideration of training requirements, performance metrics, and quality system integration. Each role must be clearly defined in terms of its contribution to product quality, with specific attention to critical control points and decision-making authority. The documentation of these requirements must balance the need for clarity and completeness with the flexibility to accommodate process improvements and technological advancements.

6.6.3 Competency Requirements and Assessment

The establishment of competency requirements represents a critical aspect of workplace management in pharmaceutical manufacturing. These requirements must reflect both the technical skills necessary for specific operations and the broader understanding of GMP principles that supports quality-oriented decision-making. The development of competency frameworks requires input from multiple stakeholders, including quality assurance, production management, and regulatory affairs, ensuring comprehensive coverage of all relevant aspects of pharmaceutical manufacturing operations.

The assessment of competency in GMP environments requires a systematic approach that combines theoretical knowledge evaluation with practical skills assessment. This comprehensive evaluation process must be documented and maintained as part of the quality system, with regular reviews to ensure continued relevance and effectiveness. The implementation of competency assessment programs requires careful consideration of both initial qualification requirements and ongoing monitoring of performance (Table 6.5).

TABLE 6.5
Competency Categories, Assessment Methods, and Documentation Requirements in GMP Environments

Competency Category	Assessment Methods	Frequency	Documentation Requirements
Technical skills	Practical demonstration	Semiannual	Detailed observation records
GMP knowledge	Written assessment	Annual	Test results and analysis
Quality systems	Interactive evaluation	Quarterly	Performance reviews
Safety protocols	Mixed assessment	Biannual	Compliance certificates

6.7 HEALTH MONITORING AND OCCUPATIONAL HEALTH SAFETY

6.7.1 Health Monitoring Programs in Pharmaceutical Manufacturing

The implementation of comprehensive health monitoring programs represents a fundamental requirement in pharmaceutical manufacturing environments. These programs must address not only the basic occupational health requirements common to all industrial settings but also the specific challenges associated with pharmaceutical manufacturing operations. The development of effective health monitoring systems requires careful consideration of exposure risks, including both acute and chronic health hazards associated with pharmaceutical manufacturing processes.

Health monitoring in pharmaceutical manufacturing extends beyond routine medical surveillance to encompass specialized monitoring requirements for different manufacturing areas. This includes specific considerations for personnel working in high-potency compound areas, biological manufacturing suites, and sterile manufacturing environments. The development of area-specific health monitoring protocols must consider both the protection of personnel and the prevention of product contamination.

The implementation of health monitoring programs requires careful attention to data collection, analysis, and trending. This includes the establishment of baseline health parameters, regular monitoring of potential exposure indicators, and prompt investigation of any adverse health trends. The management of health monitoring data must balance the need for comprehensive documentation with privacy considerations and regulatory requirements for personal health information protection.

6.7.2 Occupational Safety Systems and Controls

Occupational safety in pharmaceutical manufacturing environments requires a sophisticated approach that integrates engineering controls, administrative procedures, and personal protective equipment. The development of effective safety systems begins with comprehensive risk assessment, identifying potential hazards associated with both routine operations and exceptional circumstances. This assessment must consider not only immediate safety risks but also long-term occupational health concerns.

The implementation of engineering controls represents the primary line of defense in pharmaceutical manufacturing safety systems. These controls include containment systems, ventilation design, ergonomic equipment selection, and facility layout optimization. The integration of these elements requires careful consideration of both operational requirements and maintenance accessibility, ensuring sustainable safety performance over time.

Administrative controls provide critical support to engineering systems, establishing procedures and practices that promote safe operation while maintaining GMP compliance. These controls encompass standard operating procedures, training programs, documentation requirements, and emergency response protocols. The development of effective administrative controls requires input from multiple stakeholders, ensuring comprehensive coverage of all relevant safety aspects.

6.7.3 Emergency Response and Crisis Management

The development of emergency response capabilities represents a critical aspect of occupational health and safety in pharmaceutical manufacturing. These systems must address both routine incidents and potential crisis situations, establishing clear protocols for response, communication, and recovery. The implementation of effective emergency response systems requires regular training, simulation exercises, and continuous improvement based on experience and emerging best practices.

Crisis management in pharmaceutical manufacturing environments requires careful consideration of both immediate response requirements and long-term business continuity implications. The development of crisis management protocols must address not only safety and health concerns but also product quality implications and regulatory compliance requirements. This comprehensive approach ensures effective response to emergencies while maintaining GMP compliance and product quality standards.

6.8 TRAINING AND FUNCTIONS OWNERS SUBJECT TO PUBLIC LAW

6.8.1 Comprehensive Training Framework Implementation

The implementation of a robust training framework within pharmaceutical manufacturing organizations represents a critical component of Good Manufacturing Practice (GMP) compliance. The development and execution of training programs must encompass a wide range of considerations, from basic GMP principles to highly specialized technical skills required for specific manufacturing operations. Organizations must establish comprehensive training protocols that not only address the immediate needs of personnel but also anticipate future requirements as manufacturing technologies and regulatory expectations continue to evolve. The effectiveness of training programs depends heavily on the organization's ability to create a learning environment that promotes continuous improvement and knowledge retention while maintaining strict compliance with regulatory requirements.

Training methodologies in pharmaceutical manufacturing have evolved significantly with the integration of modern educational technologies and pedagogical approaches. Contemporary training programs utilize a blend of traditional classroom instruction, hands-on practical training, and advanced simulation technologies to ensure effective knowledge transfer. The incorporation of virtual reality and augmented reality systems has revolutionized the way operators learn complex procedures, allowing them to practice critical operations in a risk-free environment before performing them in actual manufacturing situations. This technological integration must be carefully balanced with traditional training methods to ensure comprehensive skill development and thorough understanding of GMP principles.

Documentation and assessment of training effectiveness represent crucial aspects of the training framework. Organizations must develop sophisticated systems for tracking training completion, evaluating comprehension, and maintaining detailed records of all training activities. These systems must be capable of demonstrating to regulatory authorities that personnel possess the necessary knowledge and skills to perform their assigned functions. The assessment methodology must incorporate both theoretical knowledge evaluation and practical skill demonstration, ensuring that personnel can apply their training effectively in real-world manufacturing situations.

6.8.2 Role-Specific Training Requirements

The development of role-specific training programs requires careful consideration of job functions, regulatory requirements, and organizational needs. Different positions within the pharmaceutical manufacturing environment demand varying levels of technical knowledge and GMP understanding. Quality control analysts, for example, require extensive training in analytical techniques and laboratory GMP, while production operators need focused training on specific manufacturing processes and equipment operation. The training framework must accommodate these diverse requirements while maintaining consistency in fundamental GMP principles across all roles.

Cross-functional training plays an increasingly important role in modern pharmaceutical manufacturing operations. Personnel who understand multiple aspects of the manufacturing process can contribute more effectively to problem-solving and process improvement initiatives. Organizations must carefully balance the benefits of cross-training with the need to maintain expertise in specific areas. The development of cross-training programs requires careful consideration of regulatory requirements, operational needs, and resource constraints.

6.8.3 Public Law Compliance in Training

Training programs in pharmaceutical manufacturing must align with various public law requirements, including occupational safety regulations, environmental protection standards, and specific pharmaceutical industry regulations. Organizations must ensure that their training programs address all applicable legal requirements while maintaining practical relevance to daily operations. This includes training on handling hazardous materials, environmental protection procedures, and emergency response protocols.

The intersection of public law requirements and GMP training necessitates a sophisticated approach to program development and implementation. Organizations must regularly review and update their training programs to reflect changes in legislation and regulatory requirements. This ongoing maintenance of training programs requires dedicated resources and expertise in both regulatory compliance and educational methodology (Table 6.6).

TABLE 6.6
Regulatory Training Aspects, Implementation Methods, and Assessment Frequencies in Pharmaceutical Manufacturing

Training Aspect	Regulatory Requirement	Implementation Method	Assessment Frequency
GMP basics	Mandatory annual	Classroom/online	Quarterly
Safety protocols	Initial + biannual	Practical demo	Monthly
Process specific	Role-dependent	Hands-on	Per process change
Documentation	Continuous	Workshop	Biannual

6.8.4 Continuous Professional Development

The concept of continuous professional development extends beyond basic training requirements to encompass ongoing education and skill enhancement. Organizations must create opportunities for personnel to expand their knowledge and capabilities through advanced training programs, professional certifications, and participation in industry conferences and seminars. This commitment to continuous development helps maintain a highly skilled workforce capable of adapting to evolving manufacturing technologies and regulatory requirements.

6.9 OFFICIAL REQUIREMENTS FOR PREMISES

6.9.1 Regulatory Framework for Pharmaceutical Premises

The establishment and maintenance of pharmaceutical manufacturing premises must comply with a complex framework of regulatory requirements designed to ensure product quality and safety. These requirements encompass various aspects of facility design, construction, and operation, from basic building specifications to highly specialized cleanroom environments. Organizations must navigate multiple layers of regulatory oversight, including local building codes, national GMP requirements, and international standards for pharmaceutical manufacturing facilities.

6.9.2 Structural Requirements and Building Specifications

The fundamental structural requirements for pharmaceutical manufacturing facilities demand careful consideration of materials, construction methods, and finishing details. Walls, floors, and ceilings must be constructed using materials that can withstand regular cleaning and sanitization while maintaining their structural integrity. The selection of construction materials must account for their compatibility with cleaning agents, resistance to microbial growth, and ability to maintain required environmental conditions.

Surface finishes in pharmaceutical manufacturing areas require special attention to detail and must meet stringent requirements for cleanability and durability.

Smooth, nonporous surfaces that prevent the accumulation of contamination and facilitate effective cleaning are essential. The implementation of rounded corners and coved bases helps eliminate areas where contamination might accumulate and ensures effective cleaning procedures.

6.9.3 Environmental Control Systems

The design and implementation of environmental control systems represent a critical aspect of pharmaceutical facility requirements. Heating, Ventilation, and Air Conditioning (HVAC) systems must maintain appropriate temperature, humidity, and pressure differentials while ensuring adequate air changes and filtration. The complexity of these systems increases with the requirements for different manufacturing areas, from general production spaces to specialized clean rooms.

Modern pharmaceutical facilities must incorporate sophisticated monitoring systems to ensure continuous compliance with environmental requirements. These systems must provide real-time data on critical environmental parameters and maintain detailed records for regulatory compliance. The integration of building automation systems with environmental monitoring creates a comprehensive approach to facility management and control.

6.9.4 Utility Systems and Infrastructure

Pharmaceutical manufacturing facilities require carefully designed utility systems that meet both operational needs and regulatory requirements. Water systems, in particular, demand special attention, with different quality grades required for various manufacturing operations. The design and validation of water systems must ensure consistent delivery of water meeting appropriate quality specifications while preventing contamination and maintaining system integrity.

6.9.5 Maintenance and Qualification Requirements

The maintenance of pharmaceutical manufacturing premises requires a systematic approach that ensures continuous compliance with regulatory requirements while minimizing impact on production operations. Organizations must develop comprehensive maintenance programs that address both preventive and corrective maintenance needs while maintaining appropriate documentation of all activities. The qualification and requalification of premises and associated systems represent ongoing requirements that must be carefully managed to ensure continuous compliance.

6.10 MATERIAL AND PERSONNEL FLOW AND LAYOUT

6.10.1 Principles of Material Flow Design

The design and implementation of material flow systems in pharmaceutical manufacturing facilities represents one of the most critical aspects of GMP compliance and operational efficiency. Material flow patterns must be carefully orchestrated to prevent cross-contamination, maintain product integrity, and ensure efficient manufacturing processes. The fundamental principle underlying effective material flow design centers on the concept of unidirectional flow, which minimizes the risk of mix-ups and contamination while optimizing operational efficiency. This approach requires careful consideration of facility layout, personnel movement patterns, and the integration of various manufacturing operations within the broader facility design.

In modern pharmaceutical manufacturing facilities, the implementation of material flow systems extends beyond simple directional considerations to encompass sophisticated tracking and monitoring systems. These systems must account for the movement of raw materials, intermediate products, finished goods, and waste materials throughout the facility. The complexity of material flow management is further compounded by the need to maintain appropriate environmental conditions throughout the manufacturing process, requiring careful integration of HVAC systems, airlocks, and material transfer points. This integration must consider not only the physical movement of materials but also the maintenance of appropriate pressure differentials and air quality parameters at each stage of the manufacturing process.

The establishment of effective material flow patterns requires careful consideration of regulatory requirements, operational efficiency, and risk management principles. Manufacturing facilities must implement systems that ensure complete traceability of materials while maintaining the integrity of clean and controlled environments. This includes the development of comprehensive documentation systems that track material movement, environmental conditions, and personnel interactions throughout the manufacturing process. The implementation of such systems requires careful balance between operational efficiency and regulatory compliance, with particular attention paid to critical control points where materials may be exposed to potential contamination risks (Figure 6.3).

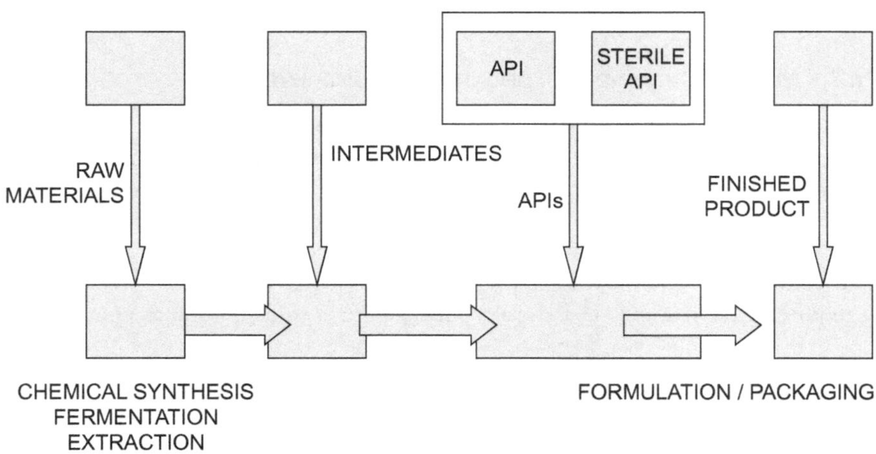

FIGURE 6.3 Integrated material and personnel flow diagram in modern pharmaceutical manufacturing.

6.10.2 Personnel Movement Optimization

The optimization of personnel movement within pharmaceutical manufacturing facilities represents a complex challenge that directly impacts both operational efficiency and product quality. Personnel flow patterns must be designed to support efficient operations while minimizing the risk of contamination and maintaining appropriate segregation between different manufacturing areas. This requires careful consideration of changing rooms, airlocks, and transition spaces that allow personnel to move between areas of different classifications while maintaining appropriate environmental controls.

The design of personnel flow patterns must account for both routine operations and emergency scenarios, ensuring that staff can efficiently access all necessary areas while maintaining appropriate containment and contamination control measures. This includes the implementation of appropriate gowning protocols, access control systems, and monitoring mechanisms that ensure compliance with GMP requirements. The complexity of personnel movement patterns is further increased by the need to accommodate different types of personnel, including manufacturing operators, quality control staff, maintenance personnel, and visitors, each with their own specific requirements and access needs.

6.10.3 Layout Optimization and Integration

The integration of material and personnel flow patterns represents a critical aspect of facility design that requires careful consideration of multiple competing factors. Modern pharmaceutical manufacturing facilities must implement layouts that support efficient operations while maintaining appropriate segregation between different manufacturing areas and activities. This includes the careful placement of production equipment, storage areas, and support facilities to minimize travel distances and optimize material handling operations while maintaining appropriate environmental controls and contamination prevention measures.

The optimization of facility layout must consider both current operational requirements and future flexibility needs, allowing for potential expansion or modification of manufacturing processes without compromising the integrity of established flow patterns. This requires careful planning and implementation of modular design concepts that can accommodate changes in product mix, manufacturing technologies, and regulatory requirements while maintaining appropriate segregation and containment measures.

6.11 AIR CLEANLINESS CLASSES AND GRADES

6.11.1 Fundamentals of Air Cleanliness Classification

The establishment and maintenance of appropriate air cleanliness standards represents a fundamental aspect of pharmaceutical manufacturing facility design and operation. Air cleanliness classifications must be carefully defined and maintained to ensure product quality and regulatory compliance. The classification system

TABLE 6.7
Air Cleanliness Classification Requirements

Grade	Maximum Permitted Number of Particles/m³
A	3,520
B	3,520
C	352,000
D	3,520,000

encompasses multiple grades of cleanliness, each with specific requirements for particle counts, microbial contamination levels, and environmental parameters. These classifications must be maintained through careful integration of HVAC systems, filtration technologies, and monitoring protocols.

Traditional air cleanliness classifications in pharmaceutical manufacturing facilities follow internationally recognized standards that define specific requirements for particle concentrations at rest and in operation. These standards provide the framework for designing and operating clean environments that support various manufacturing operations. The implementation of appropriate air cleanliness standards requires careful consideration of multiple factors, including product requirements, operational needs, and regulatory expectations (Table 6.7).

6.11.2 Design Considerations for Clean Environments

The design of clean environments in pharmaceutical manufacturing facilities requires careful consideration of multiple factors that influence air quality and cleanliness levels. This includes the selection and implementation of appropriate HVAC systems, filtration technologies, and monitoring systems that ensure consistent maintenance of required environmental conditions. The design must account for both routine operations and potential deviations, incorporating appropriate alarm systems and response protocols that maintain product quality and regulatory compliance.

The implementation of effective clean environment design requires careful integration of multiple building systems and components, including air handling units, filtration systems, and monitoring devices. These systems must work in concert to maintain appropriate pressure differentials, air change rates, and particle control measures throughout the facility. The design must also consider the impact of personnel and material movement on environmental conditions, incorporating appropriate transition spaces and control measures that minimize disruption to clean environments.

6.11.3 Monitoring and Maintenance of Air Cleanliness

The maintenance of appropriate air cleanliness levels requires implementation of comprehensive monitoring and control systems that ensure consistent compliance with established standards. This includes regular monitoring of particle counts, microbial contamination levels, and environmental parameters throughout the facility. The monitoring system must provide real-time data on environmental conditions, allowing for prompt identification and resolution of potential deviations from established parameters.

The implementation of effective monitoring systems requires careful consideration of sensor placement, data collection methods, and response protocols. This includes the establishment of appropriate alert and action limits that trigger specific responses when environmental conditions deviate from acceptable ranges. The monitoring system must also incorporate appropriate documentation and trending capabilities that support ongoing assessment of system performance and regulatory compliance.

6.12 CONSTRUCTION ELEMENTS AND BARRIER SYSTEMS

6.12.1 Fundamental Principles of Barrier System Design

The implementation of effective barrier systems in pharmaceutical manufacturing facilities represents a critical aspect of contamination control and product quality assurance. Modern barrier systems encompass a complex array of physical structures, technological solutions, and operational protocols designed to maintain the integrity of cleanroom environments and protect both products and personnel. The fundamental principles of barrier system design begin with the understanding of contamination risks and the establishment of appropriate control measures. These systems must account for various types of contamination, including particulate matter, microbiological contaminants, and cross-contamination between different products or processes. The design and implementation of barrier systems require careful consideration of material selection, construction techniques, and integration with existing facility infrastructure.

In pharmaceutical manufacturing environments, barrier systems serve multiple purposes beyond simple containment. They create distinct zones of cleanliness, establish pressure cascades, and facilitate the safe movement of materials and personnel between different manufacturing areas. The effectiveness of these systems depends heavily on their structural integrity, the appropriate selection of materials, and the implementation of robust maintenance protocols. Modern barrier systems incorporate advanced materials that resist degradation, minimize particle generation, and withstand regular cleaning and sanitization procedures. These materials must also comply with regulatory requirements and maintain their protective properties throughout their operational life cycle.

The integration of barrier systems with facility ventilation and air handling systems represents a critical aspect of contamination control. This integration requires careful consideration of airflow patterns, pressure differentials, and the maintenance of appropriate environmental conditions within protected spaces. Advanced barrier systems often incorporate sophisticated monitoring and control systems that provide real-time data on environmental conditions and system performance. These monitoring systems help ensure the continuous effectiveness of containment measures and facilitate rapid response to any deviations from specified parameters.

6.12.2 Material Selection and Construction Specifications

The selection of appropriate construction materials for barrier systems demands careful consideration of multiple factors including durability, cleanability, chemical resistance, and compatibility with manufacturing processes. Modern pharmaceutical facilities utilize a range of specialized materials engineered specifically for cleanroom environments. These materials must demonstrate exceptional resistance to cleaning agents and disinfectants while maintaining their structural integrity and protective properties. The surface characteristics of these materials play a crucial role in contamination control, with smooth, nonporous surfaces being essential for effective cleaning and sanitization.

Construction specifications for barrier systems must address not only the primary containment requirements but also the integration of necessary support systems and utilities. This includes provisions for lighting, power distribution, communication systems, and process utilities while maintaining the integrity of the containment barrier. The design must incorporate appropriate sealing systems at all joints and penetrations to prevent contamination and maintain pressure differentials. Modern construction techniques often utilize modular systems that facilitate installation and future modifications while ensuring consistent performance and reliability (Table 6.8).

6.12.3 Operational Considerations and Maintenance Protocols

The successful implementation of barrier systems requires comprehensive operational protocols and maintenance procedures. These protocols must address routine operations, cleaning and sanitization procedures, and emergency response scenarios. Personnel training plays a crucial role in ensuring the effective use of barrier systems, with emphasis on proper gowning procedures, material transfer protocols, and response to system alarms or failures. Regular maintenance activities must be

TABLE 6.8
Barrier System Components, Performance Requirements, and Maintenance Schedules in Pharmaceutical Facilities

Barrier System Component	Critical Parameters	Performance Requirements	Maintenance Frequency
Wall panels	Surface finish, joint design	Cleanability, air-tightness	Semiannual
HVAC integration	Pressure control, filtration	Continuous monitoring	Quarterly
Access systems	Sealing efficiency, durability	Personnel/material flow	Monthly
Monitoring systems	Data accuracy, response time	Real-time control	Weekly

carefully planned and executed to minimize disruption to manufacturing operations while ensuring the continued effectiveness of containment measures.

6.13 ARTIFICIAL INTELLIGENCE IN PERSONNEL MANAGEMENT AND PREMISES

6.13.1 Integration of AI Systems in GMP Environments

The integration of artificial intelligence in pharmaceutical manufacturing environments represents a transformative advancement in both personnel management and premises control. Modern AI systems offer unprecedented capabilities for monitoring, analyzing, and optimizing various aspects of manufacturing operations. These systems process vast amounts of data from multiple sources, including environmental monitoring systems, personnel tracking systems, and production equipment, to provide comprehensive insights into facility performance and operational efficiency. The implementation of AI-driven solutions requires careful consideration of regulatory requirements, validation protocols, and the need to maintain human oversight of critical processes.

6.13.2 Advanced Monitoring and Control Systems

Artificial intelligence enables sophisticated monitoring and control systems that enhance both personnel safety and facility performance. These systems utilize advanced sensors, machine learning algorithms, and real-time data analysis to detect potential issues before they impact product quality or personnel safety. The integration of AI-driven monitoring systems with traditional control systems creates a comprehensive network that provides unprecedented visibility into facility operations. This integration facilitates proactive maintenance, optimized resource allocation, and improved response to operational challenges.

6.13.3 Predictive Analytics and Resource Optimization

The application of artificial intelligence in resource optimization represents a significant advancement in pharmaceutical manufacturing efficiency. AI systems analyze historical data and current operational parameters to predict maintenance requirements, optimize personnel scheduling, and improve material flow through the facility. These predictive capabilities enable proactive maintenance scheduling, reduced downtime, and improved utilization of both human and material resources. The implementation of AI-driven optimization systems requires careful validation and continuous monitoring to ensure their effectiveness and compliance with regulatory requirements.

6.13.4 Personnel Training and Development Systems

Artificial intelligence has revolutionized approaches to personnel training and development in pharmaceutical manufacturing environments. Advanced AI systems facilitate personalized learning experiences, track competency development, and identify areas requiring additional training or support. These systems analyze individual performance data, learning patterns, and operational requirements to create targeted training programs that enhance both individual and organizational capabilities. The integration of AI in training systems enables continuous assessment of personnel competencies and facilitates rapid adaptation to changing operational requirements.

6.13.5 Compliance Monitoring and Documentation

The application of artificial intelligence in compliance monitoring and documentation management represents a significant advancement in GMP implementation. AI systems facilitate automated monitoring of compliance parameters, real-time detection of deviations, and comprehensive documentation of operational activities. These systems enhance the efficiency and accuracy of compliance monitoring while reducing the administrative burden on personnel. The implementation of AI-driven compliance systems requires careful validation and ongoing verification to ensure continued effectiveness and regulatory compliance.

6.13.6 Future Directions and Emerging Technologies

The continued evolution of artificial intelligence technologies presents new opportunities for enhancing pharmaceutical manufacturing operations. Emerging technologies such as advanced neural networks, natural language processing, and computer vision systems offer potential for further improvements in both personnel management and premises control. The integration of these technologies with existing systems requires careful consideration of validation requirements, regulatory compliance, and the need to maintain appropriate human oversight of critical processes. The future development of AI applications in pharmaceutical manufacturing will likely focus on enhanced predictive capabilities, improved integration with existing systems, and expanded applications in quality assurance and compliance monitoring.

6.14 AI APPLICATIONS IN OPTIMIZING HR FUNCTIONS AND PREMISES MANAGEMENT

6.14.1 Integration of AI Systems in Personnel Management

The integration of artificial intelligence systems into pharmaceutical manufacturing personnel management represents a transformative advancement in GMP compliance and operational efficiency. Modern pharmaceutical facilities face increasingly complex challenges in managing human resources while maintaining strict quality standards and regulatory compliance. Artificial intelligence systems have emerged as powerful tools for addressing these challenges, offering sophisticated solutions for personnel training, performance monitoring, and competency assessment. The implementation of AI-driven personnel management systems requires careful consideration of existing regulatory frameworks, validation requirements, and the need to maintain human oversight of critical processes.

The transformation of traditional human resource management through AI implementation begins with the fundamental processes of recruitment and selection. Advanced AI systems utilize natural language processing and machine learning algorithms to analyze candidate profiles, assess qualifications, and identify potential matches for specific roles within the pharmaceutical manufacturing environment. These systems go beyond simple keyword matching, incorporating deep learning capabilities that understand the context and relevance of experience, skills, and qualifications specific to GMP environments. The sophisticated analysis capabilities of AI systems enable human resource managers to make more informed decisions about candidate selection while maintaining compliance with regulatory requirements for personnel qualification and documentation.

Performance monitoring and assessment in pharmaceutical manufacturing environments have been revolutionized through the implementation of AI-driven analytics systems. These sophisticated platforms continuously analyze employee performance data, identifying patterns and trends that might indicate the need for additional training or process improvements. The integration of AI-based performance monitoring systems with traditional human resource management processes creates a comprehensive framework for ensuring consistent quality and compliance in pharmaceutical manufacturing operations. These systems can track multiple performance indicators simultaneously, providing real-time insights into employee capabilities and identifying potential areas for improvement before they impact product quality.

Training management and competency assessment represent another critical area where AI systems have demonstrated significant value in pharmaceutical manufacturing environments. Advanced learning management systems powered by artificial intelligence can adapt training content and delivery methods based on individual learning patterns and preferences, ensuring more effective knowledge transfer and skill development. These systems can automatically track training completion, assess comprehension through sophisticated testing algorithms, and generate comprehensive documentation for regulatory compliance purposes. The integration of virtual reality and augmented reality technologies, guided by AI systems, has created new possibilities for hands-on training in a controlled, risk-free environment.

Documentation management and compliance tracking have been significantly enhanced through the implementation of AI-driven systems. These platforms can automatically generate, organize, and maintain personnel-related documentation in accordance with GMP requirements, ensuring consistent compliance with regulatory standards. The ability of AI systems to process and analyze large volumes of documentation enables more effective tracking of training records, performance assessments, and health monitoring data while reducing the administrative burden on human resource personnel.

6.14.2 AI Applications in Premises Management and Monitoring

The application of artificial intelligence in pharmaceutical manufacturing premises management represents a significant advancement in maintaining facility compliance and operational efficiency. Modern AI systems provide sophisticated capabilities for monitoring and controlling environmental conditions, managing material flow, and optimizing facility utilization. These systems integrate with existing building management systems to create a comprehensive framework for ensuring consistent compliance with GMP requirements while maximizing operational efficiency.

Environmental monitoring and control systems powered by artificial intelligence have transformed the management of critical manufacturing environments. These systems utilize advanced sensors and real-time data analysis to maintain precise control over temperature, humidity, differential pressure, and particle counts in cleanroom environments. The integration of AI-driven predictive analytics enables facilities managers to anticipate potential environmental control issues before they impact production, allowing for proactive maintenance and adjustment of critical systems (Table 6.9).

6.14.3 Integration of AI in Quality Management Systems

The integration of artificial intelligence into pharmaceutical manufacturing quality management systems represents a fundamental shift in how quality control and assurance are maintained in GMP environments. AI-driven quality management systems provide sophisticated capabilities for real-time monitoring, analysis, and control of critical quality parameters across both personnel and premises domains.

TABLE 6.9
AI Application Areas, Functions, and Implementation Considerations in Premises Management

AI Application Area	Key Functions	Benefits	Implementation Considerations
Environmental monitoring	Real-time data analysis, predictive analytics	Early detection of deviations, reduced downtime	Validation requirements, data integrity
Material flow optimization	Route planning, inventory management	Improved efficiency, reduced contamination risks	Integration with existing systems
Energy management	Consumption analysis, system optimization	Cost reduction, sustainability	ROI assessment, system reliability
Maintenance planning	Predictive maintenance, resource allocation	Reduced downtime, optimal resource utilization	Training requirements, documentation

These systems enable a more proactive approach to quality management, identifying potential issues before they impact product quality or regulatory compliance.

Quality risk management has been revolutionized through the implementation of AI-driven analytics platforms. These systems continuously analyze data from multiple sources, including personnel performance metrics, environmental monitoring systems, and production parameters, to identify potential quality risks and recommend appropriate mitigation strategies. The integration of machine learning algorithms enables these systems to improve their risk assessment capabilities over time, becoming more accurate and effective in identifying potential quality issues.

The implementation of AI-driven quality management systems requires careful consideration of validation requirements and regulatory compliance. These systems must demonstrate consistent reliability and accuracy in their analysis and recommendations while maintaining appropriate levels of human oversight and control. The validation of AI systems in GMP environments presents unique challenges, requiring sophisticated approaches to system testing and verification.

6.14.4 Future Trends and Developments

The future of AI applications in pharmaceutical manufacturing personnel and premises management holds significant promise for continued advancement and innovation. Emerging technologies such as quantum computing and advanced neural networks are expected to provide even more sophisticated capabilities for analyzing complex manufacturing environments and optimizing both human and facility resources. The integration of these advanced technologies will require careful consideration of regulatory requirements and validation approaches.

The development of more sophisticated AI-driven training systems represents a significant area of future growth in pharmaceutical manufacturing personnel management. These systems will likely incorporate advanced virtual reality and augmented reality capabilities, providing increasingly realistic and effective training experiences for manufacturing personnel. The integration of emotional intelligence capabilities into AI systems may also enable more effective personnel management and development programs.

Advances in sensor technology and data analytics capabilities are expected to drive continued development of AI applications in premises management. The integration of Internet of Things (IoT) devices with AI-driven analysis platforms will enable more comprehensive monitoring and control of manufacturing environments, while advances in predictive analytics will improve the ability to anticipate and prevent potential compliance issues.

6.15 CONCLUSION

The comprehensive examination of Good Manufacturing Practice (GMP) principles relating to personnel and premises management reveals the intricate interdependence between human resources and physical infrastructure in pharmaceutical manufacturing. Throughout this chapter, we have explored the multifaceted aspects of implementing and maintaining GMP compliance, demonstrating how the successful integration of personnel management and facility design serves as the cornerstone of pharmaceutical quality assurance.

The evolution of human resource management in pharmaceutical manufacturing has demonstrated the critical importance of developing and maintaining a qualified workforce capable of consistently delivering high-quality products. The sophisticated approaches to personnel qualification, training, and development discussed in this chapter highlight the necessity of viewing human capital as a strategic asset in quality assurance. Organizations that successfully implement comprehensive training programs, maintain effective health monitoring systems, and foster a culture of quality demonstrate superior performance in maintaining GMP compliance. The integration of artificial intelligence and advanced technologies in personnel management has opened new horizons for efficiency and effectiveness while simultaneously creating new challenges that must be carefully managed.

The technical requirements for pharmaceutical manufacturing premises have evolved significantly, driven by advances in manufacturing technology and increasingly stringent regulatory requirements. The detailed examination of air cleanliness classifications, material flow optimization, and barrier systems presented in this chapter underscores the complexity of modern pharmaceutical facility design and operation. The successful implementation of these technical requirements demands a sophisticated understanding of both engineering principles and regulatory requirements, coupled with the ability to maintain these systems through effective qualification and monitoring programs.

The intersection of personnel management and premises design represents a critical aspect of GMP compliance that requires careful balance and continuous attention. The relationship between human behavior and facility design, particularly evident in cleanroom operations and material handling procedures, demonstrates the need for integrated approaches to quality assurance. Organizations must recognize that neither perfect facility design nor comprehensive personnel training alone can ensure product quality; rather, it is the harmonious integration of these elements that creates a robust quality assurance system.

Looking toward the future, the pharmaceutical manufacturing landscape continues to evolve with the integration of new technologies and manufacturing approaches. The adoption of artificial intelligence in both personnel management and premises optimization represents a significant advancement that will continue to transform the industry. However, organizations must remain mindful that these technological advancements must be balanced with fundamental GMP principles and the human elements of pharmaceutical manufacturing.

The successful implementation of GMP principles in pharmaceutical manufacturing requires a holistic understanding of how personnel and premises work together to ensure product quality. Organizations must develop sophisticated systems for managing both human resources and physical infrastructure while maintaining the flexibility to adapt to changing regulatory requirements and technological advances. The principles and practices discussed in this chapter provide a foundation for developing and maintaining effective quality systems that support both regulatory compliance and operational excellence.

In conclusion, the effective management of personnel and premises in pharmaceutical manufacturing represents a complex challenge that requires careful consideration of multiple factors. Success in this endeavor demands not only technical expertise and regulatory knowledge but also a deep understanding of human factors and organizational dynamics. As the pharmaceutical industry continues to evolve, organizations must maintain their commitment to excellence in both personnel management and facility design, recognizing that these elements form the foundation of effective GMP compliance and, ultimately, the production of safe and effective pharmaceutical products.

The insights and principles presented in this chapter serve as a valuable resource for pharmaceutical professionals seeking to develop and maintain effective systems for personnel and premises management. By understanding and implementing these principles, organizations can build robust quality systems that support both regulatory compliance and operational excellence, ultimately contributing to the production of high-quality pharmaceutical products that meet the needs of patients worldwide.

7 Facilities, Equipment, and Pharmaceutical Water

7.1 INTRODUCTION

The design, operation, and maintenance of pharmaceutical manufacturing facilities represent one of the most complex challenges in modern industrial engineering. These facilities must simultaneously satisfy stringent regulatory requirements, maintain optimal production efficiency, and ensure unwavering product quality, all while adapting to rapidly evolving technological capabilities and market demands. The intricate nature of pharmaceutical manufacturing necessitates a holistic approach to facility design and management, where every component, from the basic building materials to the most sophisticated control systems, must work in perfect harmony to ensure product integrity and patient safety.

In the contemporary pharmaceutical manufacturing environment, facility design extends far beyond mere architectural considerations. It encompasses a complex web of interconnected systems, each playing a crucial role in the overall manufacturing process. The foundation of any pharmaceutical facility lies in its basic infrastructure – the walls, floors, and ceilings that must meet stringent cleanliness requirements while withstanding aggressive cleaning agents and maintaining structural integrity. These elements must be carefully selected and designed to prevent contamination, facilitate cleaning and maintenance, and support the heavy loads of manufacturing equipment.

The evolution of pharmaceutical manufacturing has led to increasingly sophisticated approaches to process control and monitoring. Modern facilities incorporate advanced automation systems that not only control individual processes but also integrate data from multiple sources to provide comprehensive oversight of the entire manufacturing operation. This integration extends to environmental monitoring systems, which continuously track critical parameters such as temperature, humidity, differential pressure, and particulate levels, ensuring that proper conditions are maintained throughout the facility.

Water systems in pharmaceutical manufacturing deserve special attention, as water is often described as the most important raw material in pharmaceutical production. The requirements for different grades of pharmaceutical water – from purified water to water for injection – necessitate complex treatment and distribution systems that must operate continuously while maintaining consistent quality. These systems must be designed with careful consideration of microbial control, chemical purity, and system maintenance while also accounting for the specific requirements of different manufacturing processes.

DOI: 10.1201/9781003624592-7

The introduction of artificial intelligence (AI) and machine learning technologies has opened new frontages in facility management and predictive maintenance. These technologies enable pharmaceutical manufacturers to move beyond traditional preventive maintenance approaches to more sophisticated predictive and prescriptive maintenance strategies. By analyzing vast amounts of operational data, AI systems can predict equipment failures before they occur, optimize maintenance schedules, and even suggest operational adjustments to improve efficiency and product quality.

The increasing complexity of modern pharmaceutical products, particularly biologics and advanced therapy medicinal products, has placed new demands on facility design and operation. These products often require specialized handling conditions, precise environmental control, and sophisticated containment strategies. The challenge of maintaining proper conditions while ensuring operator safety and product quality has led to innovations in facility design, including the development of advanced isolation technologies and sophisticated air handling systems.

As we progress through this chapter, we will explore each of these aspects in detail, providing comprehensive insights into the principles, practices, and technologies that form the backbone of modern pharmaceutical manufacturing facilities. The discussion will be grounded in practical considerations while incorporating theoretical foundations and regulatory requirements, providing readers with a thorough understanding of this critical aspect of pharmaceutical manufacturing.

7.2 FACILITY PLANNING AND MATERIALS

7.2.1 Strategic Facility Planning and Layout Considerations

The strategic planning of pharmaceutical manufacturing facilities represents a complex multidisciplinary endeavor that requires careful consideration of numerous interconnected factors. At its core, facility planning must address not only current manufacturing requirements but also anticipate future expansion needs and technological advancements that may impact production processes. The initial phase of facility planning involves comprehensive analysis of production requirements, including batch sizes, product mix, and manufacturing technologies. This analysis must account for the entire product life cycle, from raw material receipt through final product dispatch, while considering regulatory requirements, environmental conditions, and operational efficiency.

Modern pharmaceutical facility planning incorporates lean manufacturing principles while maintaining compliance with current Good Manufacturing Practice (cGMP) requirements. The integration of these seemingly contradictory objectives requires sophisticated approaches to space utilization and flow optimization. Planners must consider personnel flows, material movements, and waste streams, ensuring that cross-contamination risks are minimized while maintaining operational efficiency. This often involves the implementation of unidirectional flow patterns and the establishment of clearly defined clean and dirty corridors.

The consideration of clean room classifications and containment requirements plays a pivotal role in facility layout decisions. Different products and processes may require varying levels of environmental control, necessitating careful planning of air handling systems and pressure cascades. The relationship between different operational areas must be carefully mapped to ensure that higher-grade areas are properly protected from potential contamination sources while maintaining efficient material and personnel flow patterns.

7.2.2 Material Selection Criteria and Implementation

The selection of construction materials for pharmaceutical facilities demands rigorous evaluation of multiple parameters including chemical resistance, cleanability, durability, and cost-effectiveness. Wall and floor materials must withstand aggressive cleaning agents while maintaining their structural integrity and appearance over extended periods. The selection process must consider not only initial installation costs but also long-term maintenance requirements and potential replacement needs (Table 7.1).

7.2.3 Environmental Control Systems Integration

Environmental control systems in pharmaceutical facilities must maintain precise conditions while operating efficiently and reliably. The design of these systems requires careful consideration of heat loads, personnel occupancy, equipment operation, and seasonal variations in external conditions. Modern facilities increasingly incorporate advanced building management systems that optimize energy usage while maintaining required environmental parameters.

The integration of environmental monitoring systems with facility control systems enables real-time tracking of critical parameters and rapid response to deviations. These systems must be designed with appropriate redundancy and backup capabilities to ensure continuous operation even during equipment failures or maintenance activities. The monitoring system architecture must support data integrity requirements while providing accessibility for operational oversight and regulatory compliance documentation.

TABLE 7.1
Critical Material Selection Parameters for Pharmaceutical Facilities

Parameter	Requirements	Evaluation Criteria	Impact on Operations
Chemical resistance	High resistance to cleaning agents	Material degradation testing	Maintenance frequency
Surface finish	Smooth, nonporous	Roughness measurements	Cleanability
Durability	Extended service life	Wear resistance testing	Life cycle cost
Installation	Minimal joints/seams	Installation complexity	Construction time
Maintenance	Easy to repair/clean	Maintenance requirements	Operational efficiency

7.2.4 Utility Systems Planning and Distribution

Utility systems form the backbone of pharmaceutical manufacturing facilities, providing essential services including electrical power, purified water, compressed air, and process gases. The planning of utility systems must ensure reliable supply while incorporating sufficient redundancy to prevent production interruptions. Distribution systems must be designed to maintain proper segregation between different grades of utilities while providing accessibility for maintenance and modification.

7.2.5 Waste Management and Environmental Considerations

Modern pharmaceutical facilities must incorporate comprehensive waste management systems that address liquid, solid, and airborne waste streams. These systems must comply with increasingly stringent environmental regulations while supporting efficient facility operation. The design of waste handling systems must consider both routine operational waste and potential upset conditions, incorporating appropriate containment and treatment capabilities.

7.3 HYGIENIC DESIGN IN SOLIDS HANDLING

7.3.1 Fundamental Principles of Hygienic Design

Hygienic design in pharmaceutical solids handling represents a critical aspect of facility design that directly impacts product quality and operational efficiency. The fundamental principles of hygienic design extend beyond simple cleanability to encompass all aspects of equipment and facility design that could affect product quality. This includes consideration of material contact surfaces, product containment, dust control, and cleaning validation requirements.

The implementation of hygienic design principles requires careful attention to detail at every level, from the selection of surface finishes to the design of equipment interfaces. Each component must be evaluated not only for its individual characteristics but also for its interaction with adjacent systems and its impact on the overall manufacturing process. The elimination of dead spaces, proper drainage design, and accessibility for cleaning and inspection represent critical considerations in hygienic design implementation.

7.3.2 Material Contact Surfaces and Finish Requirements

Surface finish requirements for product contact surfaces must balance multiple considerations including cleanability, wear resistance, and cost-effectiveness. The specification of surface finishes must consider both mechanical and electropolished surfaces, with appropriate verification methods to ensure consistent quality. The relationship between surface finish and product retention must be carefully evaluated, particularly for cohesive materials that may tend to adhere to equipment surfaces.

7.3.3 Equipment Design for Containment and Cleaning

The design of solids handling equipment must incorporate features that facilitate both containment during operation and effective cleaning between products. This includes consideration of seal designs, interface connections, and access points for cleaning and inspection. Modern equipment designs increasingly incorporate clean-in-place capabilities, requiring careful attention to spray coverage patterns and cleaning solution distribution.

7.3.4 Dust Control and Operator Protection

Effective dust control represents a critical aspect of solids handling facility design, impacting both product quality and operator safety. The design of dust control systems must consider both local extraction at specific operation points and general room air management. The selection of dust collection equipment must account for material properties, explosion risks, and cleaning requirements.

7.3.5 Validation Considerations in Hygienic Design

The validation of hygienic design features requires comprehensive testing protocols that demonstrate both initial capability and ongoing performance. This includes evaluation of cleaning procedures, containment effectiveness, and cross-contamination controls. The validation program must consider worst-case conditions and establish appropriate acceptance criteria based on product quality requirements and regulatory standards.

7.3.6 Integration of Monitoring Systems

Modern hygienic design incorporates integrated monitoring systems that provide real-time feedback on critical parameters including differential pressure, particle counts, and environmental conditions. These systems must be designed to support both routine operation and investigation of potential issues, with appropriate data storage and retrieval capabilities to support regulatory compliance requirements.

7.4 SYSTEM CONTROLLERS AND PROCESS CONTROL SYSTEMS

7.4.1 Fundamentals of Pharmaceutical Process Control

The implementation of robust process control systems in pharmaceutical manufacturing facilities represents a cornerstone of modern production methodology. These sophisticated systems integrate multiple layers of control architecture, from basic regulatory control loops to advanced supervisory systems, ensuring consistent product quality while maintaining operational efficiency. The fundamental architecture of pharmaceutical process control systems encompasses distributed control systems (DCSs), programmable logic controllers (PLCs), and supervisory control and data acquisition (SCADA) systems, working in harmony to maintain precise control over critical process parameters. These systems must operate within the framework of current Good Manufacturing Practices (cGMPs) while providing the flexibility necessary to accommodate varying production requirements and regulatory standards across different global markets.

Modern pharmaceutical process control systems incorporate multiple layers of redundancy and fault tolerance mechanisms to ensure continuous operation under various conditions. The primary control layer typically consists of robust PLC systems that handle time-critical operations and safety interlocks. These systems operate in real time, processing thousands of input/output signals per second while maintaining deterministic response times. The supervisory layer, often implemented through SCADA or DCS platforms, provides higher-level process management capabilities, including recipe management, batch control, and process optimization algorithms. This hierarchical approach to control system architecture ensures that critical process parameters remain within specified limits while allowing for sophisticated optimization and management strategies (Figure 7.1).

7.4.2 Data Acquisition and Management Systems

The acquisition, processing, and management of process data form a critical component of modern pharmaceutical manufacturing operations. Contemporary data acquisition systems must handle an ever-increasing volume of process data while ensuring data integrity and compliance with regulatory requirements. These systems

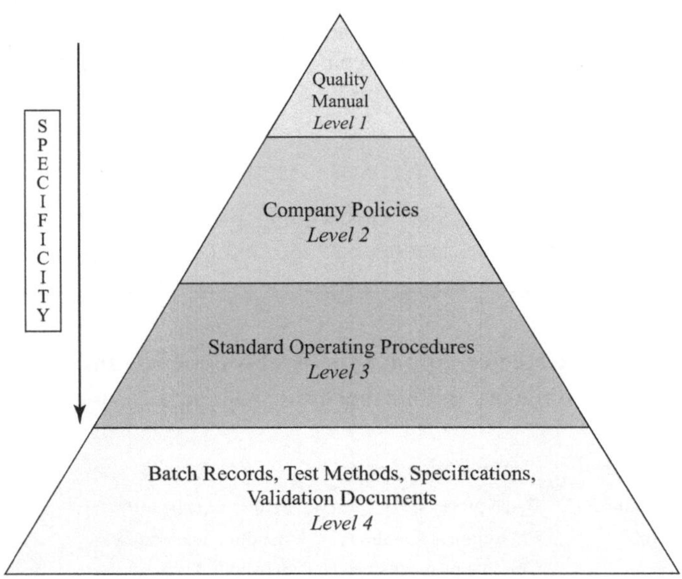

FIGURE 7.1 Hierarchical structure of a modern pharmaceutical process control system.

incorporate sophisticated data historians that capture and archive process data at high frequencies, enabling detailed analysis of process performance and facilitating regulatory compliance. The implementation of data acquisition systems requires careful consideration of sampling rates, data compression algorithms, and storage requirements to ensure optimal system performance while maintaining data accuracy and accessibility.

Modern pharmaceutical facilities employ sophisticated data management strategies that extend beyond simple data collection and storage. These systems incorporate advanced data analytics capabilities, enabling real-time process monitoring and optimization. The integration of process analytical technology (PAT) initiatives has further expanded the scope of data acquisition systems, requiring them to handle complex analytical data streams alongside traditional process parameters. The implementation of these systems necessitates careful consideration of network architecture, data security, and system scalability to ensure robust operation under varying conditions.

7.4.3 Advanced Control Strategies and Optimization

The implementation of advanced control strategies in pharmaceutical manufacturing represents a critical element in achieving consistent product quality while maintaining operational efficiency. Modern control systems incorporate sophisticated algorithms that go beyond traditional proportional-integral-derivative (PID) control, implementing model predictive control (MPC), fuzzy logic, and adaptive control strategies to handle complex process interactions. These advanced control strategies enable improved process performance through better disturbance rejection and enhanced setpoint tracking capabilities. The selection and implementation of appropriate control strategies require careful consideration of process dynamics, control objectives, and operational constraints (Table 7.2).

7.4.4 System Integration and Network Architecture

The integration of various control systems within a pharmaceutical facility requires careful consideration of network architecture and communication protocols. Modern

TABLE 7.2
Advanced Control Strategies, Application Areas, Benefits, and Implementation Complexity in Pharmaceutical Manufacturing

Control Strategy Type	Application Areas	Key Benefits	Implementation Complexity
Model predictive control	Batch processes	Predictive capability	High
Fuzzy logic control	Environmental systems	Handling uncertainty	Medium
Adaptive control	Continuous processes	Self-tuning	High
Cascade control	Temperature control	Improved response	Medium

facilities implement segregated networks that separate process control, business systems, and building management systems while maintaining necessary data exchange capabilities. The implementation of industrial ethernet protocols, such as EtherNet/IP and Profinet, enables high-speed communication between various system components while ensuring deterministic performance for critical control applications. Network security considerations play a crucial role in system design, requiring implementation of defense-in-depth strategies to protect against cyber threats while maintaining system accessibility.

7.5 TECHNICAL DOCUMENTATION, CALIBRATION, AND MAINTENANCE

7.5.1 Documentation Management Systems

The management of technical documentation in pharmaceutical facilities represents a critical aspect of regulatory compliance and operational excellence. Modern documentation management systems must handle a wide range of document types, from standard operating procedures to equipment specifications and validation protocols. These systems implement sophisticated version control mechanisms and approval workflows to ensure document integrity and compliance with regulatory requirements. The implementation of electronic document management systems has revolutionized how pharmaceutical facilities handle documentation, enabling improved accessibility while maintaining strict control over document revisions and approvals.

7.5.2 Calibration Management and Procedures

Calibration management in pharmaceutical facilities encompasses a comprehensive system of procedures, schedules, and documentation requirements designed to ensure measurement accuracy and traceability. Modern calibration management systems incorporate sophisticated scheduling algorithms that optimize calibration intervals based on instrument performance history and criticality. These systems maintain detailed records of calibration procedures, including as-found and as-left conditions, enabling trend analysis and predictive maintenance capabilities. The implementation of electronic calibration management systems has improved efficiency while ensuring compliance with regulatory requirements through automated workflow management and electronic signatures.

7.5.3 Maintenance Planning and Execution

The planning and execution of maintenance activities in pharmaceutical facilities require a systematic approach that balances equipment reliability with operational efficiency. Modern maintenance management systems implement sophisticated planning algorithms that optimize maintenance schedules based on equipment criticality, production schedules, and resource availability. These systems incorporate predictive maintenance capabilities through analysis of equipment performance data, enabling early detection of potential failures and optimization of maintenance intervals. The implementation of electronic maintenance management systems has improved maintenance efficiency through better resource allocation and improved documentation management.

7.5.4 Validation Documentation and Change Control

The documentation of validation activities and management of change control processes represent critical aspects of pharmaceutical facility operation. Modern validation documentation systems must handle complex validation protocols while ensuring traceability and compliance with regulatory requirements. These systems implement sophisticated workflow management capabilities that ensure proper review and approval of validation documentation while maintaining audit trails of all system interactions. The implementation of electronic validation documentation systems has improved efficiency through automated workflow management and improved accessibility of validation records.

7.5.5 Equipment Life Cycle Documentation

The management of equipment life cycle documentation requires comprehensive systems that track equipment from specification and procurement through installation, operation, and eventual decommissioning. Modern equipment life cycle management systems maintain detailed records of equipment specifications, maintenance history, and performance data, enabling informed decision-making regarding equipment replacement and upgrades. These systems implement sophisticated document management capabilities that ensure proper control over equipment documentation while maintaining accessibility for authorized personnel. The implementation of electronic equipment life cycle management systems has improved efficiency through better organization and accessibility of equipment documentation.

7.6 CLEANING OF FACILITIES AND CONTAINMENT IN SOLIDS HANDLING

7.6.1 Fundamentals of Pharmaceutical Facility Cleaning

The cleaning of pharmaceutical manufacturing facilities represents a critical aspect of Good Manufacturing Practice (GMP) and plays a fundamental role in preventing cross-contamination while ensuring product quality and safety. Modern pharmaceutical facility cleaning encompasses a complex array of procedures, methodologies, and validation requirements that must be carefully orchestrated to maintain the highest standards of cleanliness and sterility. The implementation of effective cleaning protocols requires a deep understanding of various factors, including material compatibility, cleaning agent selection, contact time requirements, and the specific challenges posed by different pharmaceutical products and processes.

In pharmaceutical manufacturing environments, cleaning processes must address multiple layers of contamination risk, ranging from visible residues to microscopic particles and molecular-level contamination. The selection of cleaning methodologies must take into account not only the nature of the contaminants but also the characteristics of the surfaces being cleaned, the environmental requirements of different manufacturing zones, and the potential impact on subsequent manufacturing operations. This complexity necessitates a systematic approach to cleaning validation that incorporates both analytical testing and visual inspection protocols.

The development of cleaning procedures must consider the specific requirements of different manufacturing areas, with particular attention paid to critical process equipment and direct product contact surfaces. These procedures must be validated through a comprehensive program that demonstrates both the consistency and effectiveness of the cleaning process. Documentation requirements for cleaning validation include detailed protocols, acceptance criteria, and scientific rationale for the selected cleaning parameters and analytical methods.

7.6.2 Advanced Cleaning Technologies and Methodologies

The evolution of pharmaceutical manufacturing has led to the development of sophisticated cleaning technologies that combine mechanical action, chemical cleaning agents, and controlled application methods. Automated cleaning systems, including clean-in-place (CIP) and wash-in-place (WIP) systems, have become increasingly prevalent in modern pharmaceutical facilities. These systems offer numerous advantages, including reproducibility, reduced operator exposure to cleaning agents, and improved cleaning effectiveness through precise control of critical parameters such as temperature, concentration, and contact time.

The integration of automated cleaning systems with facility management systems enables real-time monitoring and documentation of cleaning processes, providing a comprehensive record of cleaning operations and supporting data-driven optimization of cleaning procedures. Advanced monitoring technologies, including conductivity sensors, TOC analyzers, and particle counters, provide immediate feedback on cleaning effectiveness and support continuous improvement of cleaning protocols.

Contemporary cleaning validation approaches incorporate risk assessment methodologies to determine appropriate sampling locations, acceptance criteria, and testing frequencies. The implementation of quality by design (QbD) principles in cleaning validation helps ensure that cleaning processes are robust and capable of consistently meeting predetermined quality attributes. This approach requires a thorough understanding of the relationship between critical process parameters and cleaning effectiveness.

7.6.3 Containment Strategies for Solids Handling

Containment in pharmaceutical solids handling operations presents unique challenges that require carefully designed solutions to protect both product quality and operator safety. The implementation of effective containment strategies must address multiple aspects, including facility design, equipment selection, operational procedures, and monitoring systems. Modern containment approaches utilize a combination of engineering controls, administrative controls, and personal protective equipment to achieve the required level of protection.

The design of containment systems must consider the physical properties of the materials being handled, including particle size distribution, flow characteristics, and potential for static charge accumulation. The selection of appropriate containment technologies depends on various factors, including the potency of the active pharmaceutical ingredients, required handling operations, and the scale of manufacturing operations.

Advanced containment technologies incorporate features such as rapid transfer ports, split butterfly valves, and continuous liner systems to maintain containment during material transfer operations. The integration of these technologies with automated material handling systems helps minimize operator exposure while maintaining production efficiency. Monitoring systems, including real-time particle counters and pressure differential sensors, provide continuous feedback on containment effectiveness.

7.6.4 Environmental Monitoring and Control

Environmental monitoring plays a crucial role in maintaining appropriate conditions within pharmaceutical manufacturing facilities and verifying the effectiveness of cleaning and containment systems. A comprehensive environmental monitoring program must address various parameters, including particulate levels, microbial contamination, temperature, humidity, and differential pressure. The design of monitoring programs must consider both routine monitoring requirements and specialized monitoring needed during cleaning operations and containment verification.

Real-time environmental monitoring systems provide immediate feedback on facility conditions and support rapid response to any deviations from acceptable parameters. The integration of environmental monitoring data with facility management systems enables trending analysis and predictive maintenance approaches. Modern monitoring systems incorporate wireless sensors and Internet of Things (IoT) technologies to provide comprehensive coverage while minimizing installation complexity.

7.7 PHARMACEUTICAL WATER: QUALITIES, GENERATION, DISTRIBUTION, AND STORAGE

7.7.1 Water Quality Requirements in Pharmaceutical Manufacturing

Pharmaceutical water represents one of the most critical utilities in pharmaceutical manufacturing, serving as both a raw material and a cleaning agent. The requirements for pharmaceutical-grade water are precisely defined by various pharmacopoeias and regulatory agencies, with different grades of water specified for different applications. Understanding and maintaining these quality requirements demands a comprehensive approach to water system design, operation, and monitoring.

TABLE 7.3
Pharmaceutical Water Quality Requirements

Parameter	Purified Water	Water for Injection
Conductivity	<1.1 μS/cm	<1.1 μS/cm
TOC	<500 ppb	<500 ppb
Microbial limits	<100 CFU/mL	<10 CFU/100 mL
Endotoxin limits	Not specified	<0.25 EU/mL

Water quality specifications encompass both chemical and microbiological parameters, with requirements becoming increasingly stringent as water quality grades progress from purified water to water for injection (WFI). The management of water quality involves multiple unit operations, sophisticated monitoring systems, and rigorous testing protocols to ensure consistent compliance with specifications (Table 7.3).

7.7.2 Water Generation Systems

The generation of pharmaceutical-grade water involves a series of treatment steps designed to remove various impurities and achieve the required water quality specifications. Modern water generation systems incorporate multiple unit operations, including pretreatment, primary purification, and polishing steps. The design of water generation systems must consider factors such as feed water quality, required production capacity, and specific quality requirements for different applications.

Water pretreatment systems typically include multimedia filtration, water softening, and carbon filtration to remove suspended solids, hardness minerals, and chlorine. Primary purification may utilize technologies such as reverse osmosis, deionization, and ultrafiltration to achieve the required level of purification. The selection and sizing of treatment technologies must consider both current and future capacity requirements while maintaining system efficiency and reliability.

7.7.3 Water Distribution and Storage Systems

The distribution and storage of pharmaceutical water present unique challenges related to maintaining water quality and preventing microbial growth. Distribution system design must address factors such as flow rates, temperature control, and materials of construction to ensure consistent water quality at all points of use. Modern distribution systems incorporate features such as continuous circulation, point-of-use filtration, and sanitization capabilities to maintain water quality throughout the system.

Storage systems for pharmaceutical water must be designed to prevent contamination while maintaining appropriate inventory levels to meet production requirements. Tank design considerations include materials of construction, vent filtration, spray ball coverage for cleaning and sanitization, and instrumentation for monitoring critical parameters. The integration of storage systems with distribution loops requires careful consideration of hydraulic design to maintain proper flow patterns and prevent stagnation.

7.8 QUALIFICATION AND OPERATION OF WATER SUPPLIES

7.8.1 Fundamental Principles of Water System Qualification

The qualification of pharmaceutical water systems represents one of the most critical aspects of pharmaceutical facility validation. This complex process encompasses multiple stages of verification and testing, designed to ensure that the water system consistently produces water of the required quality while maintaining the system's integrity over extended periods of operation. The qualification process begins with a comprehensive risk assessment that evaluates all potential factors that could impact water quality, including source water variations, seasonal changes, and potential contamination points throughout the distribution system. This assessment forms the foundation for developing a robust qualification protocol that addresses all critical aspects of system performance and reliability.

Water system qualification requires a thorough understanding of the interplay between various system components and their potential impact on water quality. The process involves extensive documentation of system design specifications, operational parameters, and maintenance procedures. These documents must demonstrate how the system design addresses potential risks identified during the risk assessment phase and how operational controls maintain system performance within specified parameters. The qualification team must consider not only the immediate performance requirements but also long-term reliability and maintainability aspects of the system (Figure 7.2).

7.8.2 Installation Qualification Requirements

Installation qualification (IQ) for pharmaceutical water systems demands meticulous attention to detail and comprehensive documentation of all system components. The process begins with a thorough review of all installation documentation, including as-built drawings, material certificates, welding logs, and passivation records. Each component must be verified against design specifications, ensuring that materials of construction meet pharmaceutical-grade requirements and that installation methods comply with current Good Manufacturing Practices (cGMPs).

The installation qualification phase must address the physical installation aspects and the supporting documentation requirements. This includes verification of pipe slopes, dead legs, instrument calibration records, and the presence of required sampling points. Special attention must be given to sanitary connections, ensuring that they meet design specifications and are installed in a manner that prevents microbial ingress and biofilm formation. The documentation package must include detailed

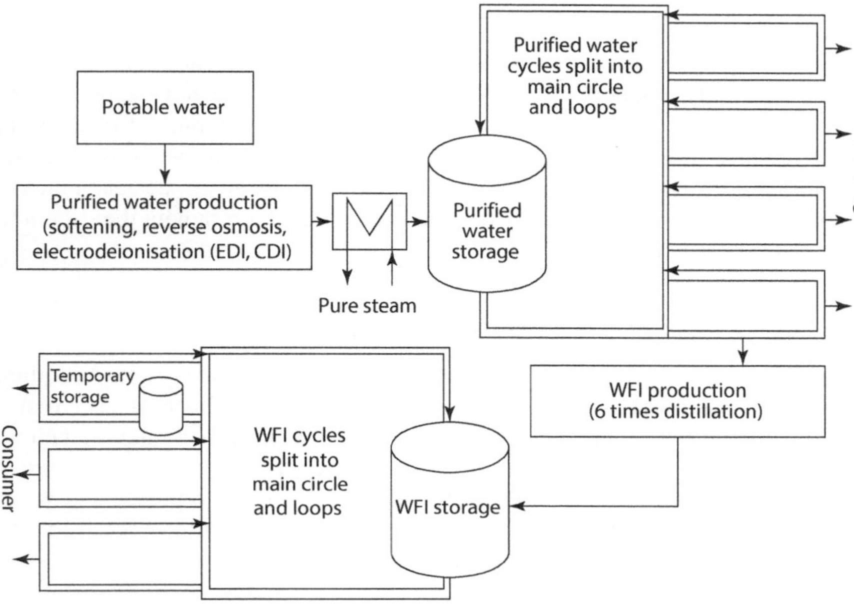

FIGURE 7.2 Comprehensive water system qualification process flow.

photographic evidence of critical installations, particularly those that will be hidden by insulation or otherwise inaccessible during normal operation.

7.8.3 Operational Qualification Protocols

Operational qualification (OQ) represents the critical phase where system functionality is verified under various operating conditions. This phase involves extensive testing of all system components, control systems, and alarms under normal and worst-case scenarios. The testing protocol must challenge the system's ability to maintain water quality under various operating conditions, including peak demand, minimum flow situations, and during system sanitization cycles (Table 7.4).

TABLE 7.4
Operational Qualification Parameters, Acceptance Criteria, and Critical Limits for Water Systems

Test Parameter	Acceptance Criteria	Testing Frequency	Critical Limits
Total organic carbon	≤500 ppb	Daily	Alert: 400 ppb
Conductivity	≤1.1 μS/cm at 20°C	Continuous	Action: 1.0 μS/cm
Bioburden	≤10 CFU/100mL	Weekly	Alert: 7 CFU/100mL
Endotoxin	≤0.25 EU/mL	Daily	Action: 0.20 EU/mL

7.8.4 Performance Qualification Strategy

Performance qualification (PQ) establishes documented evidence that the water system consistently produces water meeting all quality specifications under routine operating conditions. This phase typically extends over a minimum of one year to capture seasonal variations in source water quality and system performance. The PQ phase must include comprehensive monitoring of chemical, physical, and microbiological parameters, with sampling locations distributed throughout the system to demonstrate consistent water quality at all points of use.

7.8.5 Continuous System Monitoring

Continuous monitoring represents a critical component of water system operation and control. Modern water systems incorporate multiple online instruments measuring critical parameters such as conductivity, total organic carbon, and oxidation–reduction potential. These instruments must be integrated into a comprehensive monitoring system that provides real-time data on system performance and water quality. The monitoring strategy must include alert and action limits for each parameter, with clearly defined response protocols when limits are exceeded.

7.8.6 Preventive Maintenance Programs

A robust preventive maintenance program is essential for maintaining water system performance and compliance. This program must address all critical system components, including pumps, heat exchangers, filters, and instrumentation. Maintenance schedules must be based on manufacturer recommendations, system performance data, and risk assessment results. The program should include detailed procedures for component inspection, calibration, and replacement, with clear documentation requirements for all maintenance activities.

7.9 PURE STEAM SYSTEMS

7.9.1 Design Principles for Pure Steam Generation

Pure steam systems represent a critical utility in pharmaceutical manufacturing, particularly in sterilization applications and the production of water for injection (WFI). The design of pure steam systems requires careful consideration of materials, component selection, and system configuration to ensure the consistent production of steam meeting pharmacopoeial requirements. The system design must address thermal cycling, condensate removal, and the prevention of noncondensable gas accumulation, all while maintaining the purity of the steam produced.

7.9.2 Steam Quality Requirements

Steam quality specifications for pharmaceutical applications are stringent and must be consistently maintained. The steam must be free from additives, with controlled levels of noncondensable gases and minimal superheating. The quality parameters include chemical purity, dryness fraction, and superheat temperature, all of which must be monitored and controlled within specified limits.

7.9.3 Distribution System Design

Pure steam distribution systems require careful attention to pipe routing, insulation, and condensate management. The distribution system must maintain steam quality while preventing the formation of condensate pockets that could compromise system performance. Special consideration must be given to the design of steam traps, isolation valves, and point-of-use connections to ensure system integrity and steam quality at all use points.

7.9.4 Validation Requirements

The validation of pure steam systems encompasses multiple aspects of system performance and reliability. This includes verification of steam quality parameters, distribution system performance, and sterilization effectiveness at all points of use. The validation protocol must address both physical parameters, such as temperature and pressure, and chemical parameters, including conductivity and endotoxin levels in condensate.

7.9.5 Operational Controls and Monitoring

Effective operation of pure steam systems requires comprehensive monitoring and control strategies. This includes continuous monitoring of critical parameters such as pressure, temperature, and conductivity, along with periodic testing of condensate quality and noncondensable gas levels. The control system must maintain proper operating conditions while preventing conditions that could compromise steam quality or system integrity.

7.9.6 Maintenance and System Performance Review

Regular maintenance and performance review are essential for ensuring continued compliance and reliability of pure steam systems. This includes inspection and maintenance of critical components such as heat exchangers, steam traps, and instrumentation. The performance review process must include trend analysis of operational data, evaluation of any deviations or excursions, and assessment of system reliability and efficiency.

7.10 ARTIFICIAL INTELLIGENCE IN FACILITY MANAGEMENT

7.10.1 Integration of AI Systems in Pharmaceutical Facilities

The integration of artificial intelligence into pharmaceutical facility management represents a paradigm shift in how modern manufacturing environments are operated and maintained. Contemporary pharmaceutical facilities face increasingly complex operational challenges, from maintaining strict environmental conditions to ensuring consistent product quality across multiple production lines. Artificial intelligence systems have emerged as powerful tools for addressing these challenges, offering capabilities that extend far beyond traditional automation and control systems. These advanced systems leverage sophisticated algorithms and machine learning techniques to process vast amounts of operational data, identifying patterns and relationships that would be impossible for human operators to discern through conventional analysis methods.

The implementation of AI systems in pharmaceutical facilities begins with the establishment of a robust data collection infrastructure. This infrastructure typically encompasses thousands of sensors and monitoring points throughout the facility, collecting data on everything from environmental conditions to equipment performance metrics. These sensors form the foundation of what is often referred to as the facility's digital nervous system, continuously gathering information that feeds into central AI processing systems. The complexity of this data collection network necessitates careful planning and implementation, with consideration given to data quality, sampling rates, and the integration of legacy systems that may not have been designed with AI capabilities in mind.

Modern AI systems in pharmaceutical facilities operate on multiple levels simultaneously, from real-time process control to long-term strategic planning. At the operational level, AI algorithms continuously monitor and adjust process parameters to maintain optimal conditions, often making hundreds or thousands of minor adjustments per hour that would be impractical for human operators to manage. These systems utilize advanced machine learning techniques such as neural networks and deep learning to develop sophisticated models of facility operations, enabling them to predict and respond to changing conditions with unprecedented accuracy and speed.

The implementation of AI in facility management extends beyond mere process control to encompass broader operational optimization. AI systems analyze historical operational data to identify opportunities for efficiency improvements, from optimizing energy consumption patterns to streamlining material flow through the facility. These systems can detect subtle correlations between various operational parameters and key performance indicators, enabling facilities to optimize their operations in ways that were previously impossible. The continuous learning capabilities of modern AI systems mean that they become increasingly effective over time, constantly refining their models and adjustments based on accumulated operational data.

7.10.2 Environmental Control and Energy Optimization

AI systems have revolutionized environmental control in pharmaceutical facilities, introducing unprecedented levels of precision and efficiency in maintaining critical environmental parameters. Traditional environmental control systems operated on relatively simple feedback loops, adjusting heating, cooling, and ventilation systems based on fixed setpoints and basic control algorithms. Modern AI-driven systems, by contrast, employ sophisticated predictive models that anticipate environmental changes and adjust systems proactively rather than reactively. This predictive capability allows facilities to maintain more stable environmental conditions while simultaneously reducing energy consumption and equipment wear.

Energy optimization represents one of the most significant applications of AI in facility management. Advanced AI systems analyze patterns in energy consumption across all facility systems, from Heating, Ventilation, and Air Conditioning (HVAC) equipment to production machinery, identifying opportunities for efficiency improvements that would be difficult or impossible to detect through conventional analysis. These systems consider multiple variables simultaneously, including production schedules, weather conditions, utility rate structures, and equipment maintenance schedules, to optimize energy usage patterns across the entire facility. The sophistication of these optimization algorithms allows facilities to achieve substantial reductions in energy consumption while maintaining or improving operational performance.

7.10.3 Quality Control and Compliance Management

The application of AI in quality control and compliance management has transformed how pharmaceutical facilities maintain and document their adherence to regulatory requirements. AI systems continuously monitor thousands of operational parameters, automatically detecting deviations from established specifications and initiating appropriate responses before these deviations can impact product quality. These systems maintain comprehensive digital records of all operational parameters and events, creating an audit trail that is both more detailed and more accessible than traditional paper-based documentation systems.

The integration of AI into compliance management systems has particularly significant implications for regulatory reporting and auditing processes. AI systems can automatically generate compliance reports, highlighting potential areas of concern and providing detailed documentation of corrective actions taken. These systems also facilitate trend analysis across extended time periods, helping facilities identify and address systematic issues before they result in compliance violations. The ability of AI systems to process and analyze vast amounts of operational data makes it possible to maintain more comprehensive compliance documentation while simultaneously reducing the administrative burden on facility staff.

7.11 USING AI FOR PREDICTIVE MAINTENANCE AND EFFICIENT FACILITY OPERATIONS

7.11.1 Implementation of Predictive Maintenance Systems

The implementation of AI-driven predictive maintenance systems represents one of the most transformative applications of artificial intelligence in pharmaceutical facility management. These sophisticated systems utilize advanced machine learning algorithms to analyze equipment performance data, identifying subtle patterns and trends that may indicate developing maintenance issues long before they would be detectable through traditional monitoring methods. This predictive capability allows facilities to move beyond conventional time-based maintenance schedules to a more sophisticated condition-based approach, where maintenance activities are scheduled based on actual equipment condition and performance trends rather than arbitrary time intervals.

Predictive maintenance systems collect and analyze data from multiple sources, including vibration sensors, thermal imaging systems, power consumption monitors, and acoustic sensors. This multimodal approach to equipment monitoring provides a more comprehensive understanding of equipment condition than would be possible through any single monitoring method. Advanced AI algorithms process this data in real time, comparing current operating parameters against historical patterns and theoretical models to identify potential issues at their earliest stages. This early detection capability allows maintenance activities to be planned and executed with minimal impact on production schedules, significantly reducing unplanned downtime and associated costs.

7.11.2 Operational Efficiency Enhancement

The application of AI in operational efficiency enhancement extends far beyond traditional optimization approaches, encompassing everything from production scheduling to resource allocation. AI systems analyze historical production data alongside current operational parameters to identify opportunities for efficiency improvements across all aspects of facility operations. These systems consider multiple variables simultaneously, including equipment availability, maintenance requirements, material availability, and workforce scheduling, to optimize overall facility performance.

The sophisticated analytics capabilities of modern AI systems enable them to identify complex relationships between various operational parameters and key performance indicators. This analysis often reveals optimization opportunities that would be difficult or impossible to identify through conventional methods. For example, AI systems might identify subtle correlations between specific operating conditions and product quality metrics, enabling facilities to fine-tune their processes for optimal performance. The continuous learning capabilities of these systems mean that they become increasingly effective over time, constantly refining their models and recommendations based on accumulated operational data (Table 7.5).

7.11.3 Decision Support and Resource Optimization

Modern pharmaceutical facilities utilize AI-powered decision support systems to assist facility managers and operators in making complex operational decisions. These systems analyze vast amounts of operational data in real time, providing recommendations for everything from maintenance scheduling to resource allocation. The sophistication of these systems allows them to consider multiple competing

TABLE 7.5
AI Implementation Benefits in Pharmaceutical Facilities

Aspect	Traditional Approach	AI-Enhanced Approach	Improvement Metrics
Maintenance planning	Time based	Condition based	30%–50% reduction in maintenance costs
Energy management	Fixed schedules	Dynamic optimization	15%–25% energy savings
Quality control	Sampling based	Continuous monitoring	40%–60% reduction in quality deviations

priorities simultaneously, helping facilities balance competing demands for resources while maintaining optimal operational performance.

The integration of AI into resource optimization systems has particularly significant implications for facility efficiency and sustainability. These systems continuously monitor resource consumption patterns across all facility systems, from utilities to raw materials, identifying opportunities for optimization and waste reduction. The ability of AI systems to process and analyze vast amounts of operational data makes it possible to identify and implement optimization strategies that would be impractical or impossible to discover through conventional analysis methods.

7.12 CONCLUSION

The comprehensive examination of pharmaceutical facilities, equipment, and water systems presented in this chapter underscores the intricate complexity and critical importance of these elements in modern pharmaceutical manufacturing. The integration of facility design principles, material selection criteria, and advanced technological solutions demonstrates the evolution of pharmaceutical manufacturing toward increasingly sophisticated and efficient operations. Through careful consideration of hygienic design principles, implementation of robust control systems, and adoption of state-of-the-art water purification technologies, pharmaceutical manufacturers can ensure consistent product quality while maintaining operational efficiency.

The role of facility planning and materials selection has emerged as a cornerstone of pharmaceutical manufacturing excellence. The careful balance between operational efficiency, regulatory compliance, and quality assurance requirements demands a holistic approach to facility design and management. Modern pharmaceutical facilities represent complex ecosystems where every component, from basic infrastructure to sophisticated control systems, must work in harmony to ensure product integrity and patient safety. The implementation of hygienic design principles in solids handling operations has demonstrated the industry's commitment to maintaining the highest standards of product quality while protecting operator safety and environmental integrity.

The evolution of system controllers and process control systems has revolutionized pharmaceutical manufacturing operations. The integration of advanced automation technologies, coupled with sophisticated monitoring and data analysis capabilities, has enabled unprecedented levels of process control and optimization. These systems not only ensure consistent product quality but also provide valuable insights into process performance and potential areas for improvement. The maintenance of comprehensive technical documentation, coupled with rigorous calibration and maintenance programs, ensures the continued reliability and compliance of manufacturing operations.

Pharmaceutical water systems, including generation, distribution, and storage facilities, represent critical infrastructure elements that demand careful attention to design, operation, and maintenance. The various grades of pharmaceutical water, from purified water to water for injection, require sophisticated treatment and distribution systems that must operate continuously while maintaining consistent quality. The qualification and operation of water supplies, supported by robust monitoring and control systems, ensures the reliability and compliance of these critical utility systems.

The implementation of pure steam systems has further enhanced the industry's ability to maintain sterile conditions throughout the manufacturing process. These systems, designed and operated according to stringent requirements, provide a reliable source of sterilization media while supporting various manufacturing operations. The integration of pure steam systems with other facility utilities demonstrates the interconnected nature of modern pharmaceutical manufacturing operations.

The emergence of artificial intelligence and machine learning technologies in facility management represents a significant advancement in pharmaceutical manufacturing capabilities. These technologies enable predictive maintenance strategies, optimize operational parameters, and provide valuable insights into facility performance. The application of AI in facility management and monitoring systems has enhanced the industry's ability to maintain consistent product quality while improving operational efficiency.

Looking forward, the pharmaceutical manufacturing industry continues to evolve, driven by technological advancements and changing regulatory requirements. The integration of advanced technologies, including artificial intelligence and machine learning, will likely play an increasingly important role in facility management and optimization.

The increasing adoption of continuous manufacturing technologies will necessitate new approaches to facility design and operation. The integration of real-time monitoring and control systems will become even more critical as manufacturers seek to optimize process performance and ensure product quality. Environmental sustainability considerations will likely drive innovations in facility design and operation, with increased focus on energy efficiency and waste reduction.

The role of data analytics and artificial intelligence in facility management will continue to expand, enabling more sophisticated approaches to predictive maintenance and process optimization. The development of new pharmaceutical products, particularly in the biologics and advanced therapy medicinal products sectors, will drive innovations in facility design and environmental control systems.

As these trends continue to evolve, the fundamental principles of pharmaceutical facility design and operation discussed in this chapter will remain relevant while adapting to new challenges and opportunities. The successful integration of traditional pharmaceutical engineering principles with modern technological capabilities will ensure the continued advancement of pharmaceutical manufacturing capabilities while maintaining the highest standards of product quality and patient safety.

In conclusion, the field of pharmaceutical facilities, equipment, and water systems continues to evolve, driven by technological advancement and regulatory requirements. The successful implementation of modern facility design principles, coupled with sophisticated control systems and advanced technologies, enables pharmaceutical manufacturers to maintain the highest standards of product quality while optimizing operational efficiency. As the industry continues to advance, the principles and practices discussed in this chapter will provide a foundation for future developments in pharmaceutical manufacturing facilities and operations.

8 Qualification and Process Validation

8.1 INTRODUCTION

The field of qualification and process validation represents a critical intersection between theoretical frameworks and practical applications in algebraic noncommutative geometry. This chapter presents an exhaustive examination of the multifaceted aspects of qualification and validation processes, establishing their fundamental importance in ensuring system reliability, compliance, and operational excellence. In the contemporary landscape of increasing complexity and regulatory scrutiny, the need for robust qualification and validation processes has become paramount, driving the development of sophisticated methodologies and frameworks that address both traditional challenges and emerging requirements.

The evolution of qualification and validation processes has been marked by significant advancements in theoretical understanding and practical implementation strategies. These processes, while rooted in fundamental principles of quality assurance and regulatory compliance, have continuously adapted to incorporate new technologies, methodologies, and regulatory requirements. The integration of modern technologies, particularly artificial intelligence and automation, has introduced new dimensions to traditional validation approaches, enabling more efficient, accurate, and comprehensive validation processes (Figure 8.1).

The interplay between qualification and validation processes extends beyond mere regulatory compliance, encompassing aspects of process optimization, risk management, and continuous improvement. This comprehensive approach ensures that systems not only meet specified requirements but also operate at optimal efficiency while maintaining the highest standards of quality and reliability. The chapter explores these interconnections, highlighting how each element contributes to the overall effectiveness of validation processes.

The significance of qualification and validation processes in ensuring product quality, process consistency, and regulatory compliance cannot be overstated. These processes serve as the foundation for establishing and maintaining the integrity of systems, equipment, and procedures throughout their life cycle. Through systematic planning, execution, and documentation, qualification and validation processes provide objective evidence that systems and processes consistently produce results meeting predetermined specifications and quality attributes.

Furthermore, this chapter examines the critical role of documentation in qualification and validation processes, emphasizing the importance of maintaining comprehensive records that demonstrate compliance with established requirements

DOI: 10.1201/9781003624592-8

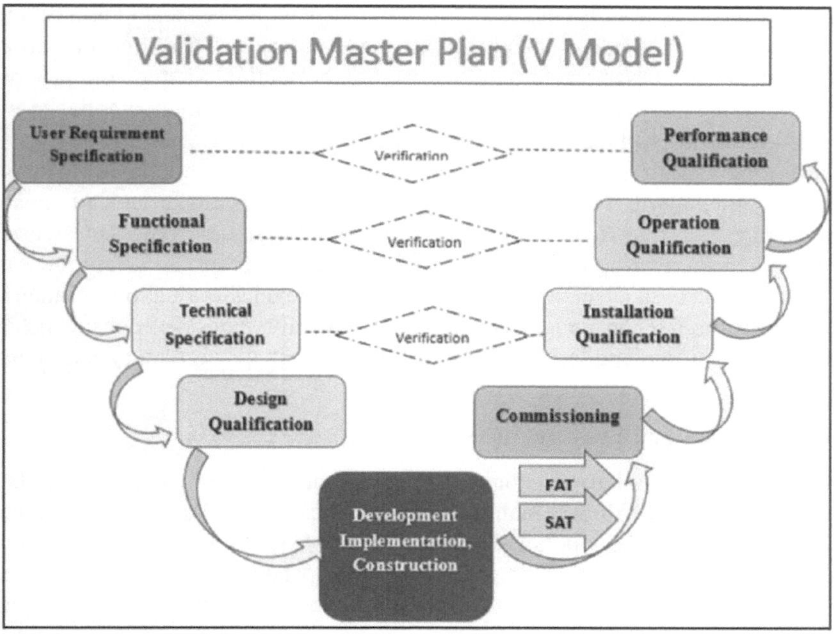

FIGURE 8.1 Evolution of qualification and validation processes through technological advancement.

and specifications. The documentation framework encompasses various elements, including protocols, reports, and standard operating procedures, forming an integral part of the overall quality management system.

This introduction sets the stage for a detailed exploration of specific aspects of qualification and validation processes, providing readers with the necessary context to understand the complexities and interrelationships of various components discussed in subsequent sections. The following sections will delve deeper into each aspect, offering comprehensive insights into requirements, methodologies, and best practices that define modern qualification and validation processes.

8.2 DESIGN QUALIFICATION (DQ), INSTALLATION QUALIFICATION (IQ), OPERATIONAL QUALIFICATION (OQ), PERFORMANCE QUALIFICATION (PQ)

8.2.1 Design Qualification (DQ)

Design qualification represents the initial and foundational phase of the qualification process, encompassing a systematic review and verification of the proposed design specifications against user requirements. This critical phase ensures that the system or equipment design aligns with intended operational parameters, regulatory requirements, and quality standards before proceeding with physical implementation. The design qualification process begins with a comprehensive assessment of user requirements, including functional specifications, performance criteria, and operational constraints. This assessment forms the basis for evaluating whether the proposed design can effectively meet these requirements while maintaining compliance with relevant regulatory standards and industry best practices.

During the design qualification phase, stakeholders conduct detailed reviews of design documentation, including engineering drawings, specifications, and control systems. These reviews focus on identifying potential design weaknesses, assessing risk factors, and ensuring that the design incorporates appropriate safety measures and quality control mechanisms. The process also includes evaluation of materials compatibility, maintenance requirements, and cleanability considerations, which are essential for ensuring long-term operational effectiveness and regulatory compliance.

8.2.2 Installation Qualification (IQ)

Installation qualification focuses on verifying that the system or equipment has been properly installed according to manufacturer specifications and design requirements. This phase encompasses a detailed verification of physical installation parameters, utility connections, and environmental conditions. The IQ process begins with a thorough review of installation documentation, including mechanical drawings, electrical schematics, and piping diagrams. Verification activities include checking physical dimensions, material specifications, and utility connections to ensure they meet design specifications and safety requirements.

A critical aspect of installation qualification involves verification of calibration requirements for instruments and control systems. This includes documenting the calibration status of all measuring devices, establishing calibration schedules, and ensuring traceability to national or international standards. The IQ phase also includes verification of safety systems, emergency stops, and alarm functions to ensure they operate as intended and provide adequate protection for operators and equipment.

8.2.3 Operational Qualification (OQ)

Operational qualification verifies that systems and equipment function according to specifications under normal operating conditions. This phase involves testing all operational parameters, control systems, and safety features to ensure they perform within established limits. The OQ process includes verification of control system functionality, including software configurations, data acquisition systems, and process control parameters. Testing activities encompass normal operating conditions as well as challenging the system at the extremes of its operational range to verify robust performance.

During operational qualification, test procedures are executed to verify system responses to various operating conditions, including start-up, shutdown, and steady-state operation. This phase also includes verification of alarm systems, interlocks, and emergency procedures to ensure they provide appropriate responses to out-of-specification conditions. Documentation of test results, including any deviations or anomalies, forms a critical part of the OQ record.

8.2.4 Performance Qualification (PQ)

Performance qualification represents the final phase of the qualification process, focusing on demonstrating that the system consistently performs according to specifications under actual operating conditions. This phase involves extended testing periods to verify consistent performance across multiple batches or operating cycles. The PQ process includes monitoring of critical process parameters, product quality attributes, and system performance metrics to establish statistical evidence of consistent operation.

Performance qualification activities typically involve running the system under normal production conditions while monitoring both process parameters and product quality attributes. This phase may include challenge testing to verify system performance under worst-case scenarios or boundary conditions. Documentation of PQ results includes statistical analysis of process capability and demonstration of consistent meeting of quality requirements.

8.3 SPECIAL CASES OF QUALIFICATION

8.3.1 Legacy System Qualification

Legacy system qualification presents unique challenges due to the age of equipment, limited documentation availability, and potential obsolescence of components. This specialized qualification approach requires careful consideration of existing documentation, historical performance data, and current regulatory requirements. The qualification process for legacy systems often involves developing new documentation to support continued use while ensuring compliance with current standards.

The qualification strategy for legacy systems typically includes a risk-based assessment to identify critical aspects requiring verification or upgrade. This may involve retrofitting modern control systems, implementing additional monitoring capabilities, or establishing enhanced maintenance programs to ensure continued reliable operation. Documentation requirements for legacy systems often necessitate creation of new standard operating procedures, maintenance protocols, and qualification protocols that reflect current operational practices while acknowledging historical performance data.

8.3.2 Computerized System Qualification

Computerized system qualification involves additional considerations related to software validation, data integrity, and electronic records management. This specialized qualification approach addresses requirements for software development life cycle documentation, system security controls, and data backup/recovery procedures. The qualification process includes verification of user access controls, audit trail functionality, and data archiving capabilities to ensure compliance with regulatory requirements for electronic systems.

8.3.3 Rapid Qualification Methodologies

Rapid qualification methodologies have emerged to address the need for accelerated validation of new technologies and processes while maintaining regulatory compliance. These approaches utilize risk-based assessments, statistical tools, and modern monitoring technologies to streamline the qualification process without compromising quality or compliance requirements. The implementation of rapid qualification methodologies requires careful planning and justification to ensure that abbreviated testing protocols provide adequate verification of system performance and reliability.

8.4 OFFICIAL REQUIREMENTS FOR VALIDATION

8.4.1 Regulatory Framework and Standards

The regulatory framework governing validation processes encompasses a complex network of requirements established by various international bodies and regulatory agencies. These requirements form the cornerstone of validation activities across different industries and applications. At the fundamental level, validation requirements are designed to ensure consistent product quality, process reliability, and compliance with established standards. The framework includes detailed specifications for documentation, testing procedures, acceptance criteria, and reporting mechanisms that must be followed throughout the validation life cycle. Regulatory bodies have established comprehensive guidelines that outline the minimum requirements for validation activities, including specific documentation requirements, testing protocols, and acceptance criteria that must be met to demonstrate compliance.

The standardization of validation requirements across different regulatory jurisdictions has led to the development of harmonized approaches that facilitate global compliance while maintaining the flexibility to address specific regional requirements. These harmonized approaches consider various aspects of validation, including risk assessment methodologies, documentation requirements, and validation strategies. The evolution of regulatory requirements has been driven by advances in technology, emerging industry practices, and lessons learned from historical validation experiences. This continuous evolution necessitates regular updates to validation procedures and documentation to ensure alignment with current regulatory expectations.

8.4.2 Documentation Requirements

The documentation requirements for validation processes represent a critical component of regulatory compliance and serve as evidence of proper validation execution. Comprehensive documentation must be maintained throughout the validation life cycle, from initial planning through execution and final approval. This documentation includes validation master plans, protocols, reports, and supporting documentation that demonstrate compliance with established requirements. The structure and content of validation documentation must follow specific guidelines established by regulatory authorities, ensuring consistency and completeness in validation activities.

Validation documentation must provide clear evidence of the rationale behind validation decisions, including risk assessments, acceptance criteria selection, and testing methodologies. The documentation should establish clear traceability between validation requirements, testing activities, and results, enabling reviewers to understand the validation process and verify compliance with established requirements. The maintenance of accurate and complete documentation throughout the validation life cycle is essential for demonstrating regulatory compliance and supporting continuous improvement initiatives.

8.5 VALIDATION PLANNING AND PROCEDURE

8.5.1 Development of Validation Master Plan

The development of a comprehensive validation master plan represents the initial step in establishing a structured approach to validation activities. This master plan serves as the foundational document that outlines the overall validation strategy, including scope, responsibilities, timelines, and resource requirements. The plan must address various aspects of validation, including equipment qualification, process validation, cleaning validation, and computer system validation, where applicable. The master plan should establish clear links between validation activities and quality system requirements, ensuring alignment with overall quality objectives.

The validation master plan must incorporate risk assessment methodologies to identify critical aspects of processes and systems that require validation. This risk-based approach helps optimize validation efforts by focusing resources on areas that have the greatest impact on product quality and process reliability. The plan should also establish clear criteria for determining the extent of validation required for different systems and processes, taking into consideration factors such as complexity, criticality, and potential impact on product quality.

8.5.2 Protocol Development and Execution

The development and execution of validation protocols represent critical steps in the validation process. Protocols must be developed with careful consideration of regulatory requirements, industry standards, and specific process characteristics. These protocols should outline detailed testing procedures, acceptance criteria, and documentation requirements for each validation activity. The protocol development process must involve subject matter experts from various disciplines to ensure comprehensive coverage of all aspects requiring validation.

Protocol execution must follow established procedures and timelines, with careful attention to documentation requirements and data integrity considerations. Any deviations from established protocols must be thoroughly documented and evaluated for potential impact on validation results. The execution phase should include regular reviews of progress and results to ensure alignment with validation objectives and regulatory requirements. This systematic approach to protocol execution helps ensure consistency and reliability in validation activities while maintaining compliance with regulatory requirements.

8.5.3 Results Analysis and Reporting

The analysis of validation results and preparation of comprehensive validation reports represent crucial elements of the validation process. Results analysis must follow established statistical methodologies where applicable, with clear documentation of analysis methods and acceptance criteria. The analysis should consider both quantitative and qualitative aspects of validation results, providing a comprehensive assessment of compliance with established requirements. Any deviations or nonconformances identified during validation must be thoroughly investigated and documented, with appropriate corrective actions implemented as needed.

Validation reports must provide a clear and complete record of validation activities, including detailed descriptions of methods used, results obtained, and conclusions reached. These reports should establish clear links between validation requirements, testing activities, and results, enabling reviewers to understand the validation process and verify compliance with established requirements. The reports must include appropriate statistical analysis where applicable, demonstrating the validity of conclusions reached during the validation process.

8.5.4 Change Management in Validation

The management of changes during and after validation represents a critical aspect of maintaining validated status. Changes to validated systems or processes must be carefully evaluated for potential impact on validation status, with appropriate revalidation activities conducted as needed. A formal change management system must be established to ensure proper evaluation, documentation, and implementation of changes affecting validated systems or processes. This system should include clear procedures for assessing change impact, determining revalidation requirements, and maintaining appropriate documentation throughout the change process.

The change management process must include appropriate risk assessment methodologies to evaluate the potential impact of changes on product quality and process reliability. Changes must be properly documented and approved before implementation, with appropriate validation activities conducted to demonstrate maintained compliance with established requirements. The change management system should also include procedures for communicating changes to affected personnel and updating relevant documentation to reflect implemented changes.

8.6 OFFICIAL REQUIREMENTS FOR QUALIFICATION

8.6.1 Regulatory Framework and Standards

The regulatory framework governing qualification requirements encompasses a comprehensive set of standards and guidelines that establish the fundamental basis for validation processes in noncommutative geometric systems. These requirements are derived from multiple regulatory bodies and international standards organizations, forming a complex web of interconnected specifications that must be carefully navigated to ensure compliance. The framework emphasizes the importance of systematic approaches to qualification, requiring organizations to establish documented evidence that their systems, equipment, and processes consistently meet predetermined specifications and quality attributes (Figure 8.2).

At the core of qualification requirements lies the concept of risk-based approach to validation, which demands a thorough understanding of potential risks associated with system implementation and operation. This approach requires organizations to conduct comprehensive risk assessments that identify potential failure modes, evaluate their impact on product quality and process reliability, and implement appropriate control measures. The risk-based methodology extends beyond mere compliance, incorporating elements of continuous improvement and process optimization that enable organizations to maintain competitive advantages while ensuring regulatory compliance.

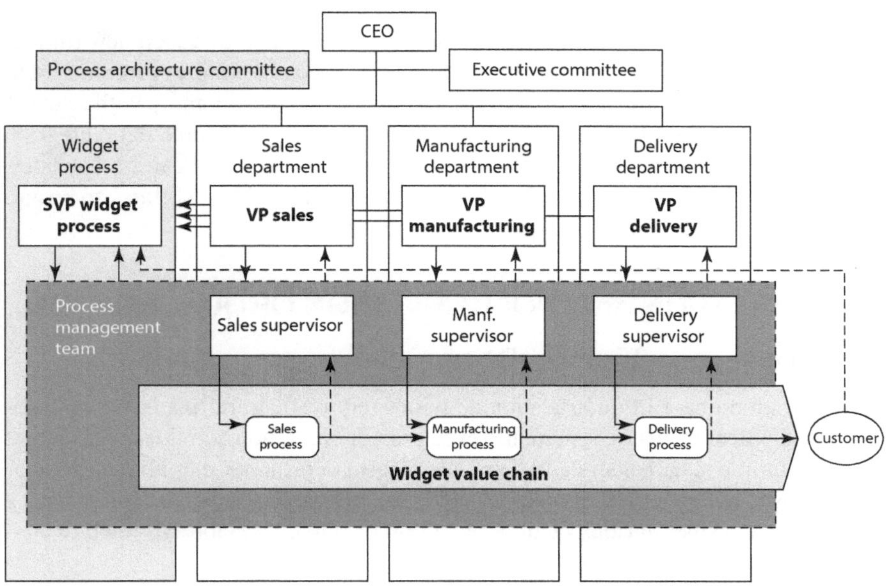

FIGURE 8.2 Hierarchical structure of regulatory requirements and their interrelationships in qualification processes.

The evolution of qualification requirements reflects the increasing complexity of modern systems and the need for more sophisticated validation approaches. Contemporary requirements emphasize the importance of integrated validation strategies that address not only individual system components but also their interactions within larger operational contexts. This holistic approach ensures that qualification activities consider the broader implications of system implementation and maintain focus on overall process effectiveness and efficiency.

8.6.2 Documentation Requirements

Documentation requirements form a critical component of qualification processes, establishing the framework for demonstrating compliance and maintaining transparency throughout the validation life cycle. The documentation structure must provide clear evidence of systematic planning, execution, and verification of qualification activities, ensuring traceability and accountability at every stage of the process. This comprehensive documentation approach serves multiple purposes, including regulatory compliance, knowledge management, and process improvement.

Proper documentation practices require organizations to maintain detailed records of qualification activities, including protocols, test results, deviations, and corrective actions. The documentation must be sufficiently detailed to allow independent verification of qualification activities while remaining clear and accessible to relevant stakeholders. This balance between comprehensiveness and clarity presents unique challenges that organizations must address through well-designed documentation systems and procedures.

The integration of electronic documentation systems has introduced new dimensions to qualification documentation requirements, necessitating appropriate controls for electronic records and signatures. These systems must comply with relevant regulations while providing improved accessibility, searchability, and data integrity. The implementation of electronic documentation systems requires careful consideration of security measures, backup procedures, and audit trail capabilities to ensure the authenticity and reliability of qualification records.

8.7 PREPARATION AND DOCUMENTATION OF QUALIFICATION

8.7.1 Planning and Protocol Development

The preparation phase of qualification activities requires careful planning and protocol development to ensure systematic execution of validation activities. This phase involves multiple stakeholders and demands careful coordination to align qualification objectives with organizational goals and regulatory requirements. The planning process must consider various factors, including resource availability, timeline constraints, and risk management strategies.

Protocol development represents a critical aspect of qualification preparation, requiring detailed specification of test procedures, acceptance criteria, and documentation requirements. Protocols must be sufficiently comprehensive to ensure consistent execution while maintaining flexibility to accommodate unexpected situations or deviations. The development process involves multiple iterations and reviews to ensure all aspects of qualification are adequately addressed.

8.7.2 Execution and Documentation Practices

The execution phase of qualification activities must follow established protocols while maintaining detailed documentation of all activities, results, and observations. This phase requires careful attention to detail and strict adherence to predetermined procedures to ensure the validity and reliability of qualification results. Documentation practices during execution must capture not only successful outcomes but also deviations, investigations, and corrective actions.

The importance of contemporaneous documentation cannot be overstated, as it provides the primary evidence of qualification activities and their outcomes. Organizations must establish robust systems for managing documentation flow, ensuring proper review and approval processes, and maintaining document control throughout the qualification life cycle. These systems must accommodate both planned activities and unexpected situations while maintaining compliance with regulatory requirements.

8.7.3 Review and Approval Processes

The review and approval of qualification documentation involves multiple layers of oversight to ensure completeness, accuracy, and compliance with predetermined requirements. This process requires participation from various stakeholders, including technical experts, quality assurance personnel, and management representatives. The review process must be systematic and thorough, with clearly defined roles and responsibilities for each participant.

Final approval of qualification documentation represents a critical milestone that signifies successful completion of validation activities and readiness for operational implementation. The approval process must include appropriate checks and balances to ensure objectivity and independence in decision-making. Organizations must maintain clear records of review and approval activities, including any comments, recommendations, or conditions associated with approval decisions.

8.8 PROCESS VALIDATION AND PRODUCT LIFE CYCLE

The concept of process validation within the product life cycle framework represents a comprehensive and systematic approach to ensuring product quality, consistency, and regulatory compliance throughout all stages of development and production. This section explores the intricate relationship between process validation and the various phases of the product life cycle, emphasizing the dynamic nature of validation activities and their critical role in maintaining product quality over time.

8.8.1 Life Cycle Approach to Process Validation

Process validation in the context of product life cycle management encompasses a holistic understanding of how validation activities evolve and adapt throughout different stages of product development and commercialization. The life cycle approach recognizes that validation is not a one-time event but rather a continuous process that begins during product development and extends throughout the entire commercial production phase. This comprehensive perspective ensures that quality is built into the product from the earliest stages of development and maintained throughout its commercial life.

During the initial stages of product development, process validation activities focus on establishing a thorough understanding of the critical process parameters and their relationships to product quality attributes. This understanding is developed through systematic experimentation, risk assessment, and the application of scientific principles. The knowledge gained during these early stages forms the foundation for subsequent validation activities and helps establish the control strategy that will be implemented during commercial production.

The integration of process validation within the product life cycle framework necessitates a strong emphasis on knowledge management and continuous improvement. Organizations must establish robust systems for capturing, analyzing, and utilizing process knowledge gained throughout the product life cycle. This knowledge management approach enables more effective decision-making regarding process changes, improvements, and optimization efforts while maintaining the validated state of the process.

8.8.2 Continuous Process Verification

Continuous process verification represents a modern approach to process validation that emphasizes the ongoing collection and analysis of process data throughout the commercial phase of the product life cycle. This approach enables organizations to maintain a state of control while identifying opportunities for process improvement and optimization. The implementation of continuous process verification requires sophisticated data collection and analysis systems, as well as established procedures, for responding to trends and addressing potential issues before they impact product quality.

Statistical process control methods play a crucial role in continuous process verification, providing the tools necessary for monitoring process performance and detecting trends that may indicate a need for intervention. These methods must be carefully selected and implemented to ensure they are appropriate for the specific process being monitored and capable of detecting meaningful changes in process performance.

8.8.3 Change Management in Validated Processes

The management of changes within validated processes represents a critical aspect of the life cycle approach to process validation. Changes may be necessary for various reasons, including process optimization, equipment upgrades, or regulatory requirements. The challenge lies in implementing these changes while maintaining the validated state of the process and ensuring continued product quality.

A robust change management system must be established to evaluate proposed changes, assess their potential impact on product quality, and determine the appropriate level of revalidation required. This system should include clear procedures for documenting changes, conducting risk assessments, and obtaining necessary approvals before implementation.

8.9 CLEANING VALIDATION: MASTER PLAN, SCOPE, ACCEPTANCE CRITERIA, SAMPLING, ANALYTICAL PROCEDURES, DOCUMENTATION

8.9.1 Development of Cleaning Validation Master Plan

The cleaning validation master plan serves as the foundational document that outlines the overall approach to cleaning validation within an organization. This comprehensive document establishes the principles, strategies, and requirements for cleaning validation activities across all applicable equipment and processes. The master plan must address various aspects, including regulatory requirements, organizational policies, and specific cleaning procedures, for different types of equipment and products.

The development of a cleaning validation master plan requires input from multiple functional areas within the organization, including quality assurance, manufacturing, engineering, and analytical services. This collaborative approach ensures that all relevant aspects of cleaning validation are adequately addressed and that the resulting plan is both comprehensive and practical for implementation.

8.9.2 Establishing Acceptance Criteria

The establishment of appropriate acceptance criteria represents a critical element of the cleaning validation process. These criteria must be scientifically sound, achievable in practice, and capable of ensuring product quality and patient safety. The determination of acceptance criteria typically involves consideration of various factors, including product toxicity, equipment surface characteristics, and analytical method capabilities.

8.9.3 Sampling Methodology and Analysis

A well-designed sampling program is essential for demonstrating the effectiveness of cleaning procedures and verifying compliance with established acceptance criteria. The sampling strategy must consider various factors, including equipment design, potential contamination locations, and the physical and chemical properties of residues. Both direct surface sampling and rinse sampling methods may be employed, depending on the specific requirements of the validation program.

8.10 ARTIFICIAL INTELLIGENCE IN VALIDATION PROCESSES

8.10.1 Introduction to AI in Validation

The integration of artificial intelligence (AI) in validation processes represents a transformative shift in how organizations approach quality assurance and compliance verification. This revolutionary advancement has fundamentally altered the traditional validation paradigms, introducing unprecedented levels of efficiency, accuracy, and predictive capabilities. The implementation of AI-driven validation processes encompasses machine learning algorithms, neural networks, and advanced data analytics, creating a sophisticated ecosystem that enhances the overall validation lifecycle while maintaining regulatory compliance and quality standards. This section explores the multifaceted applications of AI in validation processes, examining both the theoretical foundations and practical implementations that are reshaping the industry landscape.

8.10.2 AI-Driven Validation Methodologies

The implementation of AI-driven validation methodologies encompasses a comprehensive approach that leverages advanced algorithms and data processing capabilities to enhance traditional validation processes. These methodologies incorporate sophisticated pattern recognition systems, predictive analytics, and automated decision-making frameworks that significantly improve the accuracy and efficiency of validation procedures. Organizations implementing AI-driven validation systems benefit from real-time monitoring capabilities, automated anomaly detection, and predictive maintenance alerts that help prevent validation failures before they occur. The integration of machine learning algorithms enables these systems to continuously learn from historical validation data, improving their accuracy and effectiveness over time.

8.10.3 Data Analytics and Pattern Recognition

The application of advanced data analytics and pattern recognition in validation processes has revolutionized how organizations identify and address potential validation issues. These sophisticated systems analyze vast amounts of historical and real-time validation data, identifying subtle patterns and correlations that might be imperceptible to human observers. The implementation of deep learning algorithms enables these systems to understand complex relationships between various validation parameters, providing invaluable insights for process optimization and risk mitigation. This enhanced analytical capability allows organizations to move from reactive to proactive validation strategies, significantly reducing the likelihood of validation failures and compliance issues.

8.10.4 Regulatory Compliance and AI Integration

The integration of AI in validation processes necessitates careful consideration of regulatory requirements and compliance standards. Organizations must ensure that

AI-driven validation systems maintain transparency, traceability, and accountability while meeting established regulatory guidelines. This involves developing comprehensive documentation frameworks that detail the AI algorithms' decision-making processes, validation criteria, and performance metrics. The implementation of AI in regulated environments requires careful balance between innovation and compliance, ensuring that advanced technological capabilities enhance rather than compromise regulatory adherence.

8.11 AUTOMATING QUALIFICATION AND VALIDATION PROCESSES WITH AI TOOLS

8.11.1 Framework for Automation Implementation

The automation of qualification and validation processes through AI tools represents a significant advancement in quality assurance and compliance verification. This comprehensive framework encompasses the integration of intelligent automation systems, machine learning algorithms, and advanced analytics platforms to streamline and enhance traditional qualification and validation procedures. The implementation strategy focuses on creating seamless interactions between automated systems and human operators, ensuring optimal efficiency while maintaining regulatory compliance and quality standards. Organizations implementing these automated systems must carefully consider factors such as system architecture, data integrity, and process control to ensure successful integration.

8.11.2 Intelligent Process Automation

Intelligent process automation (IPA) in qualification and validation processes combines robotic process automation (RPA) with artificial intelligence capabilities to create sophisticated systems capable of handling complex validation tasks. These automated systems utilize advanced algorithms to perform routine validation procedures, analyze results, and generate comprehensive reports with minimal human intervention. The implementation of IPA systems significantly reduces the time and resources required for validation activities while improving accuracy and consistency. Organizations benefit from automated workflow management, real-time monitoring capabilities, and intelligent decision support systems that enhance the overall validation process.

8.11.3 Quality Control and Monitoring Systems

The implementation of automated quality control and monitoring systems represents a critical component of AI-driven validation processes. These sophisticated systems utilize advanced sensors, real-time data analysis, and machine learning algorithms to continuously monitor validation parameters and identify potential issues before they impact product quality or compliance. The integration of automated monitoring systems enables organizations to maintain consistent quality standards while reducing the resource requirements associated with manual monitoring processes.

8.11.4 Data Management and Analysis Automation

The automation of data management and analysis processes in validation activities encompasses comprehensive systems for data collection, storage, analysis, and reporting. These automated systems utilize advanced data processing algorithms to handle large volumes of validation data, identifying trends, patterns, and potential issues that might affect validation outcomes. The implementation of automated data management systems ensures data integrity, traceability, and compliance with regulatory requirements while significantly reducing the time and effort required for data analysis and reporting activities.

8.11.5 Validation Documentation and Reporting

The automation of validation documentation and reporting processes represents a significant advancement in maintaining comprehensive records of validation activities. These automated systems utilize intelligent document processing capabilities to generate, manage, and archive validation documentation while ensuring compliance with regulatory requirements. The implementation of automated documentation systems significantly reduces the administrative burden associated with validation activities while improving the accuracy and completeness of validation records (Table 8.1).

8.11.6 Future Directions and Challenges

The continuous evolution of AI technologies and automation capabilities presents both opportunities and challenges for organizations implementing automated validation processes. Future developments in artificial intelligence and machine learning will likely introduce more sophisticated validation capabilities, enabling even greater levels of automation and intelligence in validation processes. However, organizations must carefully consider challenges related to system validation, regulatory compliance, and change management when implementing these advanced technologies. The successful implementation of automated validation systems requires a balanced approach that leverages technological capabilities while maintaining regulatory compliance and quality standards.

TABLE 8.1
AI Tool Categories, Functions, and Regulatory Impact in Validation Documentation and Reporting

AI Tool Category	Primary Function	Implementation Complexity	Regulatory Impact
Process automation	Workflow management	High	Moderate
Data analytics	Pattern recognition	Moderate	High
Documentation	Record management	Low	High
Quality control	Monitoring	High	Critical
Decision support	Risk assessment	Moderate	Significant

8.12 CONCLUSION

The comprehensive exploration of qualification and process validation in this chapter underscores their fundamental importance in ensuring system reliability, regulatory compliance, and operational excellence within the context of algebraic noncommutative geometry. Through detailed examination of various qualification types, validation requirements, and implementation strategies, several key insights emerge that shape our understanding of modern validation practices.

The evolution of qualification processes, from traditional approaches to contemporary methodologies incorporating artificial intelligence and automation, represents a significant advancement in the field. This transformation has enabled organizations to achieve higher levels of efficiency, accuracy, and comprehensiveness in their validation activities. The integration of advanced technologies has not only streamlined validation processes but also introduced new capabilities for data analysis, pattern recognition, and predictive maintenance that were previously unattainable.

The interconnected nature of qualification and validation processes, as demonstrated throughout this chapter, emphasizes the importance of holistic approaches to system validation. The relationship between design qualification (DQ), installation qualification (IQ), operational qualification (OQ), and performance qualification (PQ) forms a comprehensive framework that ensures thorough validation at every stage of system implementation and operation. This systematic approach, combined with robust documentation practices and risk management strategies, provides organizations with the tools necessary to maintain compliance while optimizing operational efficiency.

Cleaning validation, as a specialized aspect of process validation, exemplifies the detailed attention required for specific validation activities. The comprehensive treatment of master planning, scope definition, acceptance criteria, sampling methodologies, and analytical procedures provides practitioners with practical guidance for implementing effective cleaning validation programs. The importance of scientific rigor in developing and executing cleaning validation protocols cannot be overstated, as it directly impacts product quality and patient safety.

The emergence of artificial intelligence in validation processes represents a paradigm shift in traditional validation approaches. AI-driven tools and methodologies offer unprecedented capabilities for data analysis, pattern recognition, and predictive maintenance, enabling organizations to implement more proactive and efficient validation strategies. The automation of qualification and validation processes, while presenting new challenges, offers significant opportunities for improving accuracy, consistency, and efficiency in validation activities.

Future directions in qualification and validation processes will likely continue to be shaped by technological advancements, regulatory evolution, and increasing system complexity. The need for adaptive validation strategies that can accommodate emerging technologies while maintaining regulatory compliance will drive further innovation in validation methodologies. Organizations must remain vigilant in maintaining the balance between innovation and compliance, ensuring that new approaches align with fundamental validation principles and regulatory requirements.

The critical role of documentation in validation processes remains paramount, serving as the primary evidence of compliance and system reliability. The transition to electronic documentation systems, while offering improved efficiency and accessibility, requires careful consideration of data integrity, security, and long-term preservation requirements. Organizations must continue to adapt their documentation practices to accommodate technological advancements while maintaining compliance with regulatory requirements.

In conclusion, the field of qualification and process validation continues to evolve, driven by technological advancement, regulatory requirements, and operational needs. The principles and methodologies presented in this chapter provide a solid foundation for understanding and implementing effective validation strategies. As organizations face increasingly complex validation challenges, the ability to adapt and innovate while maintaining compliance will become increasingly important. The future of qualification and validation processes lies in the successful integration of advanced technologies with established validation principles, enabling organizations to achieve higher levels of efficiency, accuracy, and reliability in their validation activities.

This comprehensive understanding of qualification and process validation, combined with practical implementation strategies and emerging technologies, positions organizations to meet current challenges while preparing for future developments in the field. The continued evolution of validation practices will require ongoing attention to regulatory requirements, technological capabilities, and operational needs, ensuring that validation processes remain effective and relevant in an ever-changing landscape.

9 Production, Sanitation, and Sterile Packaging

9.1 INTRODUCTION

The pharmaceutical manufacturing landscape represents a complex ecosystem where product quality, patient safety, and regulatory compliance converge to create a demanding operational environment. The fundamental principles of pharmaceutical production extend far beyond mere manufacturing processes, encompassing a comprehensive framework of sanitation protocols, environmental controls, and sterile packaging requirements. This intricate web of interconnected systems and procedures forms the backbone of modern pharmaceutical manufacturing, ensuring the consistent production of safe, effective, and high-quality therapeutic products.

In the contemporary pharmaceutical industry, the implementation of robust production and sanitation protocols has become increasingly sophisticated, driven by advances in technology and evolving regulatory requirements. The integration of modern automation systems, real-time monitoring capabilities, and advanced analytical tools has transformed traditional manufacturing approaches into highly controlled, data-driven operations. This transformation necessitates a thorough understanding of both fundamental principles and cutting-edge technologies, enabling pharmaceutical professionals to navigate the complex landscape of current Good Manufacturing Practices (cGMPs) while maintaining optimal production efficiency.

The evolution of pharmaceutical manufacturing has witnessed a paradigm shift from conventional batch processing to more sophisticated continuous manufacturing systems, accompanied by advanced environmental monitoring and control mechanisms. This transformation has been particularly evident in sterile manufacturing operations, where the demands for product quality and contamination control have driven the development of increasingly sophisticated production environments and technologies. The integration of artificial intelligence and machine learning algorithms has further enhanced these capabilities, enabling predictive maintenance, real-time quality control, and automated process optimization (Figure 9.1).

The complexity of modern pharmaceutical manufacturing extends to the intricate relationships between various operational aspects, including personnel training and hygiene, environmental monitoring, material handling, and quality control processes. These elements must work in harmony to ensure the consistent production of pharmaceutical products that meet stringent quality standards while maintaining operational efficiency and regulatory compliance. The implementation of comprehensive sanitization protocols, coupled with advanced sterile packaging

DOI: 10.1201/9781003624592-9

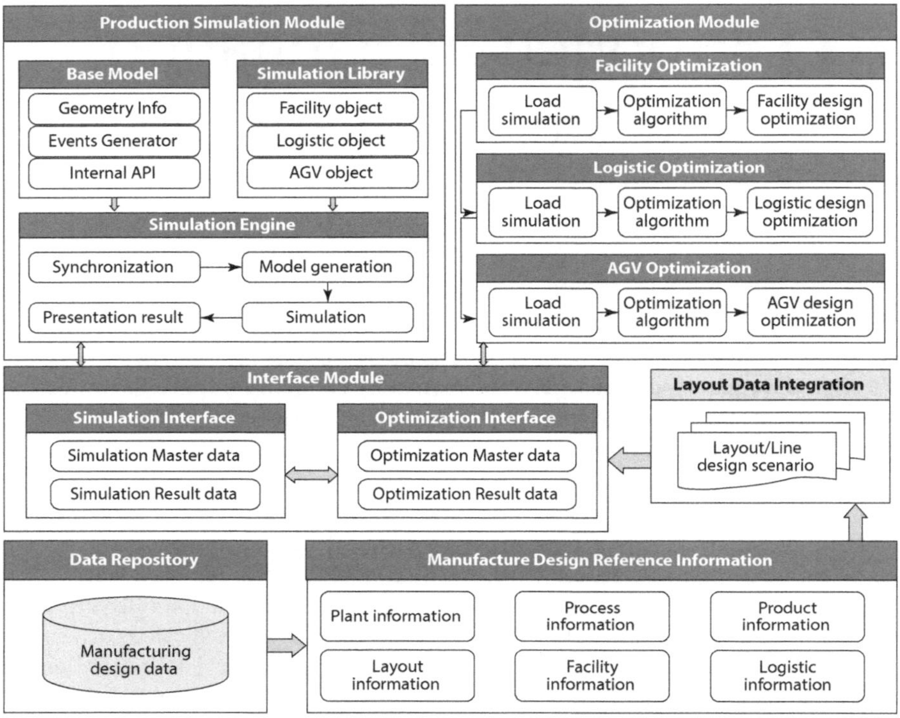

FIGURE 9.1 Modern pharmaceutical facility layout.

technologies, ensures product integrity throughout the manufacturing process and beyond.

The scope of pharmaceutical production encompasses a wide range of specialized processes and technologies, from basic material handling and processing to sophisticated sterile manufacturing operations. This chapter examines these elements in detail, providing readers with a comprehensive understanding of the principles, practices, and technologies that define modern pharmaceutical manufacturing. The integration of traditional manufacturing principles with advanced technologies and automated systems creates a dynamic environment where quality, efficiency, and compliance must be carefully balanced to achieve optimal results.

The contemporary challenges facing pharmaceutical manufacturers have driven the development of innovative solutions and technologies, particularly in the areas of contamination control, environmental monitoring, and sterile packaging. These advancements have led to the emergence of new methodologies and approaches, from advanced air handling systems and cleanroom technologies to sophisticated packaging solutions and automated inspection systems. The continuous evolution of these technologies, coupled with increasingly stringent regulatory requirements, necessitates a thorough understanding of both fundamental principles and emerging trends in pharmaceutical manufacturing.

9.2 SANITATION AND PERSONNEL HYGIENE

9.2.1 Fundamentals of Personnel Hygiene in Pharmaceutical Manufacturing

Personnel hygiene in pharmaceutical manufacturing environments represents a critical cornerstone of product quality and contamination control. The implementation of comprehensive hygiene protocols extends far beyond basic cleanliness practices, encompassing a sophisticated system of procedures, monitoring mechanisms, and behavioral guidelines designed to minimize the risk of product contamination from human sources. In modern pharmaceutical facilities, personnel hygiene programs integrate multiple layers of control, from basic hand washing procedures to complex gowning protocols for cleanroom environments. These programs must account for various factors including the movement patterns of personnel, the nature of their interactions with products and equipment, and the specific requirements of different manufacturing zones.

The establishment of effective personnel hygiene protocols begins with a thorough understanding of human-related contamination sources and their potential impact on product quality. This understanding drives the development of comprehensive training programs that address not only the technical aspects of hygiene procedures but also the underlying scientific principles and regulatory requirements. Personnel must be educated about the relationship between their actions and potential product contamination, enabling them to make informed decisions about their behavior in controlled environments. The training process must be continuous and adaptive, incorporating new requirements and best practices as they emerge while reinforcing fundamental principles through regular refresher sessions.

9.2.2 Advanced Gowning Procedures and Contamination Control

The implementation of proper gowning procedures represents a critical aspect of personnel hygiene in pharmaceutical manufacturing environments. Modern gowning protocols have evolved into sophisticated systems that incorporate multiple layers of protective clothing, specific dressing sequences, and detailed monitoring procedures to ensure compliance. These protocols must be carefully designed to balance the need for contamination control with practical considerations such as worker comfort and mobility. The selection of appropriate gowning materials and designs must take into account factors such as particle generation, durability, and compatibility with different manufacturing environments.

9.2.3 Monitoring and Documentation Systems

The effectiveness of personnel hygiene programs relies heavily on robust monitoring and documentation systems. Modern pharmaceutical facilities employ various technologies and methodologies to track compliance with hygiene protocols, from simple observational audits to sophisticated electronic monitoring systems. These systems generate valuable data that can be analyzed to identify trends, potential issues, and opportunities for improvement in hygiene practices. The documentation of hygiene-related activities must be comprehensive and readily accessible for review by quality assurance personnel and regulatory inspectors.

9.3 PRODUCTION HYGIENE AND ENVIRONMENTAL MONITORING

9.3.1 Integrated Approach to Production Environment Control

Production hygiene in pharmaceutical manufacturing encompasses a broad range of activities and controls designed to maintain the cleanliness and integrity of manufacturing environments. This includes the implementation of sophisticated cleaning and sanitization protocols, environmental monitoring programs, and contamination control strategies. Modern pharmaceutical facilities employ advanced technologies and methodologies to maintain appropriate environmental conditions, including temperature, humidity, differential pressure, and particulate levels. These controls must be carefully coordinated to ensure consistent compliance with regulatory requirements while maintaining operational efficiency.

9.3.2 Advanced Environmental Monitoring Systems

Environmental monitoring represents a critical component of production hygiene programs in pharmaceutical manufacturing. Modern monitoring systems incorporate various technologies and methodologies to track environmental parameters and detect potential contamination issues. These systems must be capable of generating real-time data about environmental conditions while maintaining accurate records for regulatory compliance purposes. The design and implementation of environmental monitoring programs require careful consideration of sampling locations, frequencies, and alert/action limits (Table 9.1).

9.3.3 Cleaning and Sanitization Protocols

The development and implementation of effective cleaning and sanitization protocols form a fundamental aspect of production hygiene. These protocols must be

TABLE 9.1
Environmental Monitoring Requirements for Different Clean Room Classifications

Parameter	ISO 5	ISO 6	ISO 7	ISO 8
Particles ≥0.5 μm/m³	3,520	35,200	352,000	3,520,000
Viable monitoring frequency	Each shift	Daily	Weekly	Monthly
Temperature range (°C)	18–22	18–22	18–25	18–25
Relative humidity (%)	45 ± 5	45 ± 5	45 ± 10	45 ± 15

carefully designed to ensure the removal of potential contaminants while preventing the introduction of cleaning residues that could affect product quality. Modern cleaning procedures incorporate validated methodologies and specialized cleaning agents designed for pharmaceutical manufacturing environments. The effectiveness of cleaning procedures must be regularly verified through appropriate testing and monitoring programs.

9.3.4 Risk Assessment and Control Strategies

Production hygiene programs must be based on comprehensive risk assessments that identify potential sources of contamination and evaluate their potential impact on product quality. These assessments inform the development of control strategies that address identified risks through appropriate preventive measures and monitoring programs. Risk assessment methodologies must be regularly updated to account for changes in manufacturing processes, regulatory requirements, and emerging best practices.

9.4 GMP IN THE PRODUCTION PROCESS

9.4.1 Implementation of GMP Principles

Good Manufacturing Practice (GMP) implementation in pharmaceutical production processes represents a cornerstone of quality assurance and regulatory compliance. The comprehensive integration of GMP principles throughout the manufacturing life cycle ensures consistent product quality while minimizing risks associated with production variability and contamination. This systematic approach encompasses every aspect of production, from raw material handling to finished product release, requiring meticulous documentation and validation at each step. The establishment of robust standard operating procedures (SOPs) forms the foundation of GMP compliance, providing detailed instructions for all critical operations and ensuring consistency across different production batches and operators.

The implementation of GMP principles extends beyond mere procedural compliance, incorporating sophisticated quality management systems that enable continuous monitoring and improvement of production processes. These systems facilitate the early detection of potential quality issues and enable prompt corrective actions, thereby maintaining the integrity of pharmaceutical products throughout their manufacturing life cycle. The integration of risk assessment methodologies within GMP frameworks allows manufacturers to identify critical control points and establish appropriate mitigation strategies, ensuring robust quality control while optimizing resource allocation and operational efficiency.

9.4.2 Documentation and Recordkeeping Requirements

Documentation within GMP systems serves as the backbone of quality assurance, providing tangible evidence of compliance with established procedures and regulatory requirements. The hierarchical structure of GMP documentation encompasses various levels of documentation, from high-level quality manuals to detailed batch records and analytical procedures. This comprehensive documentation system ensures traceability of all manufacturing operations, enabling thorough investigation of any deviations or quality issues that may arise during production.

9.4.3 Quality Risk Management in Production

The integration of quality risk management principles within GMP frameworks enables manufacturers to adopt a proactive approach to quality assurance. This systematic methodology involves the identification, assessment, and mitigation of potential risks throughout the production process. The implementation of risk-based approaches allows manufacturers to allocate resources effectively, focusing attention on critical aspects of production that have the greatest potential impact on product quality and patient safety.

9.5 WEIGH-IN, IDENTIFICATION, AND IN-PROCESS CONTROL

9.5.1 Material Weighing Systems and Procedures

The accurate weighing and dispensing of materials represent critical operations in pharmaceutical manufacturing, requiring precise equipment calibration and rigorous procedural controls. Modern weighing systems incorporate advanced technologies for error prevention and data integrity assurance, including automated verification systems and electronic data capture capabilities. The implementation of sophisticated material tracking systems ensures complete traceability of all components used in production, from raw materials to packaging components.

9.5.2 Material Identification and Verification

Material identification and verification procedures form an essential component of pharmaceutical quality assurance, preventing mix-ups and ensuring the use of correct materials in production. Advanced identification technologies, including spectroscopic methods and barcode systems, enable rapid and accurate verification of materials at various stages of the manufacturing process. The integration of electronic verification systems with production management software provides real-time tracking and documentation of material usage throughout the manufacturing facility (Table 9.2).

9.5.3 In-Process Control Strategies

In-process control strategies encompass a wide range of monitoring and testing procedures designed to ensure consistent product quality throughout the manufacturing

TABLE 9.2
Material Verification Requirements by Material Type

Material Type	Primary Verification Method	Secondary Verification	Documentation Requirements
Active ingredients	Identity testing	Visual inspection	Full analysis record
Excipients	Certificate of analysis	Identity testing	Verified CoA
Packaging materials	Dimensional check	Visual inspection	Inspection record

process. These controls include both physical measurements and analytical testing, providing real-time feedback on critical quality attributes. The implementation of automated monitoring systems enables continuous assessment of process parameters, allowing for immediate detection and correction of any deviations from established specifications.

9.5.4 Process Analytical Technology Integration

The integration of process analytical technology (PAT) represents a significant advancement in pharmaceutical manufacturing control strategies. PAT systems enable real-time monitoring of critical process parameters and quality attributes, facilitating immediate process adjustments when necessary. This advanced monitoring capability enhances process understanding and control, leading to improved product consistency and reduced manufacturing variability.

9.5.5 Data Management and Analysis

The management and analysis of in-process control data play a crucial role in maintaining product quality and process efficiency. Modern data management systems facilitate the collection, storage, and analysis of large volumes of process data, enabling trend analysis and predictive maintenance capabilities. The implementation of sophisticated statistical process control methods allows manufacturers to identify and address potential quality issues before they impact product quality.

9.5.6 Deviation Management and Corrective Actions

The establishment of robust deviation management systems ensures appropriate handling of any departures from established procedures or specifications. This comprehensive approach includes thorough investigation of root causes, implementation of effective corrective actions, and verification of their effectiveness. The integration of deviation management within the broader quality system ensures systematic handling of quality issues and continuous improvement of manufacturing processes.

9.6 PREVENTION OF CROSS-CONTAMINATION, REWORKING, WAREHOUSE, AND LOGISTICS

9.6.1 Prevention of Cross-Contamination

The prevention of cross-contamination in pharmaceutical manufacturing environments represents one of the most critical aspects of quality assurance and regulatory compliance. This complex process encompasses multiple layers of control strategies, facility design considerations, and operational protocols that work in concert to maintain product integrity throughout the manufacturing life cycle. Modern pharmaceutical facilities implement sophisticated airlocking systems, pressure cascades, and dedicated Heating, Ventilation, and Air Conditioning (HVAC) systems to create physical barriers between different production areas, effectively minimizing the risk of cross-contamination between products. These engineering controls are complemented by comprehensive standard operating procedures (SOPs) that govern material flow, personnel movement, and equipment usage across different manufacturing zones.

The implementation of effective cross-contamination prevention strategies requires a thorough understanding of product characteristics, manufacturing processes, and potential contamination pathways. Pharmaceutical manufacturers must conduct detailed risk assessments to identify critical control points where cross-contamination risks are highest, enabling the development of targeted prevention strategies. These assessments consider factors such as product potency, toxicity, and physical characteristics as well as operational aspects such as equipment sharing, material handling procedures, and cleaning validation requirements. The results of these assessments inform facility design decisions, equipment selection, and the development of cleaning and sanitization protocols (Figure 9.2).

9.6.2 Reworking Procedures and Quality Control

The reworking of pharmaceutical products presents unique challenges that require careful consideration of quality implications, regulatory requirements, and operational efficiency. Comprehensive reworking procedures must be established to ensure that any reprocessed material meets all quality specifications while maintaining compliance with regulatory guidelines. These procedures typically involve detailed documentation requirements, specific testing protocols, and rigorous quality control measures to verify the integrity of reworked materials. The decision to rework product must be based on a thorough evaluation of product quality attributes, stability considerations, and potential impacts on patient safety.

9.6.3 Warehouse Management and Material Flow

Modern pharmaceutical warehousing operations have evolved into sophisticated systems that integrate advanced technology with traditional storage and handling practices. The implementation of warehouse management systems (WMS) enables

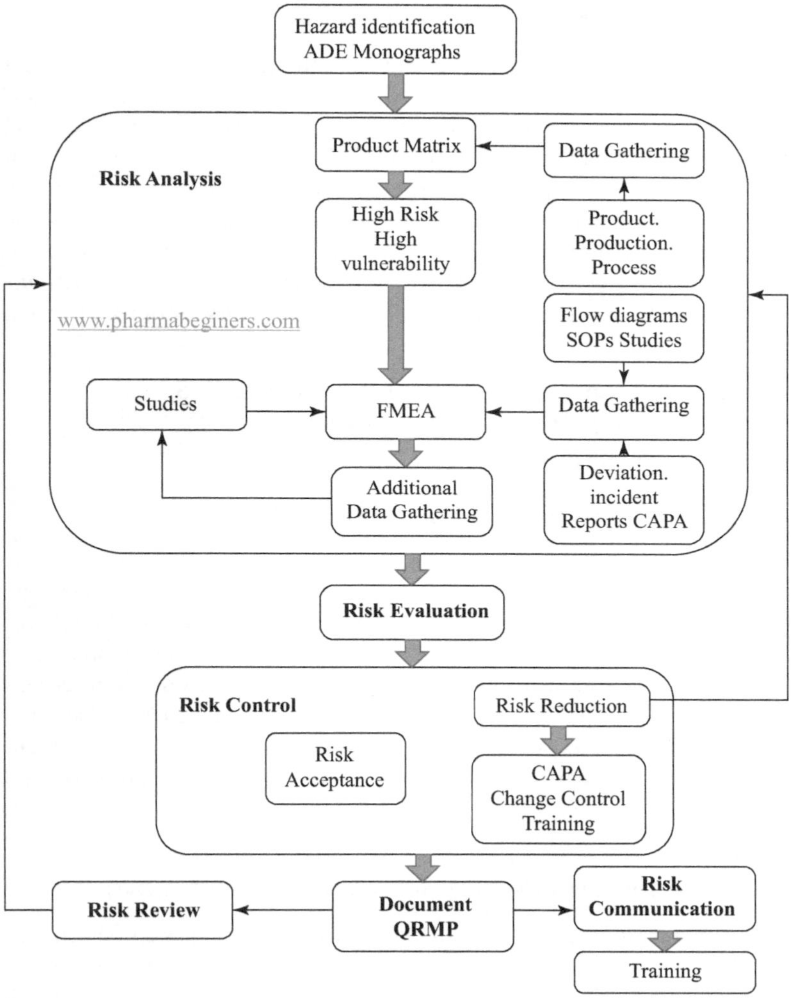

FIGURE 9.2 Multilevel cross-contamination prevention strategy framework in pharmaceutical manufacturing.

real-time tracking of inventory movements, environmental conditions, and storage requirements for different material categories. These systems must accommodate the diverse storage requirements of raw materials, intermediate products, and finished goods while maintaining appropriate segregation between different product categories. Temperature-mapped storage areas, humidity-controlled zones, and specialized handling equipment ensure that material quality is maintained throughout the storage period.

9.6.4 Logistics and Supply Chain Integration

The logistics component of pharmaceutical manufacturing operations extends beyond simple material movement to encompass a complex network of suppliers, transportation providers, and distribution channels. Modern pharmaceutical logistics systems must ensure product integrity throughout the supply chain while maintaining compliance with good distribution practice (GDP) requirements. This includes the implementation of temperature monitoring systems, chain of custody documentation, and qualification of transportation providers. The integration of track-and-trace technologies enables real-time monitoring of product movement and environmental conditions during transit, providing valuable data for quality assurance purposes.

9.7 STERILE PRODUCTION: AIR LOCK CONCEPTS, MANUFACTURE OF TERMINALLY STERILIZED PRODUCTS, STERILIZATION PROCESSES

9.7.1 Air Lock Design and Implementation

The concept of airlocks in sterile pharmaceutical manufacturing represents a fundamental aspect of contamination control and clean room design. Modern airlock systems incorporate sophisticated pressure cascade arrangements, HVAC controls, and monitoring systems to maintain appropriate environmental conditions while facilitating the movement of materials and personnel between different clean room classifications. The design of airlock systems must consider factors such as room classification requirements, material flow patterns, and personnel movement frequencies to ensure optimal functionality while maintaining contamination control effectiveness.

9.7.2 Clean Room Classification and Environmental Controls

Sterile manufacturing facilities implement rigorous environmental control systems based on clean room classification requirements defined by international standards. These systems typically incorporate High Efficiency Particulate Air (HEPA) filtration, precise temperature and humidity controls, and continuous monitoring of particulate levels and microbial contamination. The establishment of appropriate environmental conditions requires careful consideration of air change rates, pressure differentials, and filtration requirements specific to each clean room classification level (Table 9.3).

9.7.3 Terminal Sterilization Processes

Terminal sterilization represents a critical process in the manufacture of sterile pharmaceutical products, encompassing various methodologies such as steam sterilization, dry heat sterilization, and radiation sterilization. Each sterilization method presents unique advantages and limitations that must be carefully evaluated based

TABLE 9.3
Clean Room Classification Parameters and Environmental Control Requirements for Sterile Manufacturing

Clean Room Parameter	Class A/ ISO 5	Class B/ISO 6	Class C/ISO 7	Class D/ISO 8
Maximum particles ≥0.5 μm/m³	3,520	35,200	352,000	3,520,000
Maximum viable organisms/m³	<1	10	100	200
Air changes per hour	>20	>20	>20	>10
Pressure differential (Pa)	>15	>10	>5	>--

on product characteristics and quality requirements. The validation of terminal sterilization processes requires extensive documentation, including the development of detailed protocols, execution of validation studies, and ongoing monitoring of critical process parameters.

9.7.4 Process Validation and Monitoring

The validation of sterile manufacturing processes encompasses multiple elements, including equipment qualification, process validation, and ongoing monitoring of critical parameters. This comprehensive approach ensures consistent achievement of sterility assurance levels while maintaining product quality attributes. Process validation activities must consider various factors such as sterilization parameters, product loading patterns, and environmental conditions to ensure robust and reproducible results.

9.8 ASEPTIC PROCESSING AND FREEZE-DRYING

9.8.1 Principles of Aseptic Processing

The foundation of aseptic processing in pharmaceutical manufacturing represents a complex interplay of controlled environments, sophisticated equipment, and meticulously designed procedures aimed at maintaining product sterility throughout the manufacturing process. In modern pharmaceutical facilities, aseptic processing encompasses a comprehensive system of controls that begins with the design and maintenance of cleanroom environments and extends through every aspect of the manufacturing process. These environments are characterized by stringent air quality requirements, carefully controlled pressure differentials, and sophisticated filtration systems that work in concert to maintain the sterile conditions necessary for product integrity. The implementation of aseptic processing requires a thorough understanding of microbial control strategies, particle management, and environmental monitoring techniques, all of which must be integrated into a cohesive system that ensures product sterility from beginning to end.

The complexity of aseptic processing extends beyond mere environmental controls to encompass the intricate choreography of material flow, personnel movement, and equipment operation within the sterile manufacturing space. This orchestration requires careful consideration of factors such as air flow patterns, material transfer procedures, and personnel gowning protocols, all of which must be validated and continuously monitored to maintain the sterile barrier system. The development and implementation of robust standard operating procedures (SOPs) plays a crucial role in ensuring consistency and compliance across all aspects of aseptic processing, from basic material handling procedures to complex manufacturing operations.

9.8.2 Advanced Freeze-Drying Technologies

The evolution of freeze-drying technology in pharmaceutical manufacturing has led to the development of sophisticated systems that combine precise control of critical process parameters with advanced monitoring capabilities. Modern freeze-drying equipment incorporates multiple layers of control and monitoring systems, enabling precise management of temperature, pressure, and time parameters throughout the lyophilization cycle. The integration of process analytical technology (PAT) has further enhanced these capabilities, providing real-time monitoring and control of critical process parameters while enabling adaptive process control strategies that optimize cycle efficiency and product quality.

9.8.3 Process Validation and Quality Assurance

The validation of aseptic processing and freeze-drying operations represents a cornerstone of pharmaceutical quality assurance, encompassing a comprehensive evaluation of all critical process parameters and their impact on product quality. This validation process extends beyond basic equipment qualification to include detailed assessments of cleaning procedures, sterilization processes, and environmental monitoring systems. The development of robust validation protocols requires careful consideration of regulatory requirements, industry standards, and specific product characteristics, all of which must be integrated into a comprehensive validation strategy that ensures consistent product quality and compliance with regulatory requirements.

9.9 TESTING FOR STERILITY, ENDOTOXINS, LEAKAGE, AND PARTICLES

9.9.1 Comprehensive Sterility Testing Methodologies

The implementation of effective sterility testing protocols represents a critical aspect of pharmaceutical quality control, requiring the integration of multiple testing methodologies and sophisticated analytical techniques. Modern sterility testing approaches combine traditional culture-based methods with advanced rapid microbial detection systems, enabling more efficient and reliable assessment of product sterility. These testing protocols must be carefully validated to ensure their ability to detect potential contamination while maintaining the sensitivity and specificity required for regulatory compliance.

9.9.2 Endotoxin Detection and Control

The management of endotoxin contamination in pharmaceutical products requires a comprehensive approach that combines sophisticated testing methodologies with robust control strategies. Modern endotoxin testing protocols utilize advanced detection methods such as the Limulus amebocyte lysate (LAL) test and its variations, enabling precise quantification of endotoxin levels in pharmaceutical products and raw materials. The implementation of effective endotoxin control strategies requires careful consideration of multiple factors, including raw material quality, process controls, and environmental monitoring systems, all of which must work together to ensure product safety and compliance with regulatory requirements.

9.9.3 Advanced Particle Detection Systems

The evolution of particle detection technology in pharmaceutical manufacturing has led to the development of sophisticated systems that enable real-time monitoring and control of particulate contamination. Modern particle detection systems incorporate advanced optical technologies and automated analysis capabilities, providing comprehensive data on particle size distributions and concentrations throughout the manufacturing process. The integration of these systems into the overall quality control strategy requires careful consideration of sampling protocols, alert and action limits, and response procedures, all of which must be validated to ensure reliable detection and control of particulate contamination.

9.9.4 Leak Detection and Container Integrity Testing

The assurance of container closure integrity represents a critical aspect of pharmaceutical packaging validation, requiring the implementation of sophisticated testing methodologies and control strategies. Modern leak detection systems utilize advanced technologies such as high-voltage leak detection, vacuum decay, and laser-based imaging systems to ensure the integrity of primary packaging components. The development of comprehensive container integrity testing protocols requires careful consideration of multiple factors, including package design, material properties, and environmental conditions, all of which must be integrated into a robust testing strategy that ensures product stability and safety throughout its shelf life.

9.10 MICROBIOLOGICAL MONITORING

9.10.1 Introduction to Microbiological Monitoring

Microbiological monitoring represents a cornerstone of pharmaceutical quality control systems, encompassing a comprehensive array of testing procedures, sampling methodologies, and analytical techniques designed to ensure product safety and compliance with regulatory requirements. The implementation of robust microbiological monitoring programs extends beyond simple testing procedures, incorporating sophisticated sampling strategies, advanced analytical methodologies, and comprehensive data management systems. This complex network of monitoring activities serves as a critical component in maintaining product quality and patient safety throughout the pharmaceutical manufacturing process.

9.10.2 Environmental Monitoring Systems

The establishment of effective environmental monitoring systems requires careful consideration of multiple factors, including sampling locations, frequency of testing, and appropriate analytical methods. Modern pharmaceutical facilities implement sophisticated environmental monitoring programs that incorporate both passive and active air sampling techniques, surface monitoring protocols, and personnel monitoring procedures. These comprehensive monitoring systems generate significant volumes of data that must be carefully analyzed and interpreted to identify potential trends or issues before they impact product quality. The integration of automated sampling systems with advanced data management platforms has revolutionized environmental monitoring capabilities, enabling real-time detection and response to potential contamination events.

9.10.3 Sample Collection and Analysis Protocols

The development and implementation of effective sample collection protocols represent a critical aspect of microbiological monitoring programs. These protocols must address various sampling methodologies, including air sampling, surface sampling, and personnel monitoring, while ensuring that samples are collected in a manner that accurately represents the environmental conditions being monitored. The selection of appropriate sampling locations, timing, and frequencies must be based on a thorough understanding of facility operations, product characteristics, and potential contamination risks.

9.10.4 Data Management and Trend Analysis

Modern microbiological monitoring systems generate vast amounts of data that must be effectively managed and analyzed to provide meaningful insights into facility performance and potential quality issues. The implementation of sophisticated data management systems enables the integration of monitoring data from multiple sources, facilitating comprehensive trend analysis and early detection of potential problems. These systems incorporate advanced statistical analysis tools and visualization capabilities, enabling quality control personnel to identify subtle trends or patterns that might indicate emerging contamination risks.

9.10.5 Alert and Action Limits

The establishment of appropriate alert and action limits represents a critical component of effective microbiological monitoring programs. These limits must be based on historical data, regulatory requirements, and product-specific considerations while allowing for appropriate responses to potential contamination events. The development of these limits requires careful consideration of multiple factors, including facility design, operational parameters, and product characteristics, ensuring that monitoring programs can effectively identify and respond to potential quality issues before they impact product safety.

9.11 PACKAGING MATERIALS AND PROCESSES

9.11.1 Introduction to Pharmaceutical Packaging

Pharmaceutical packaging represents a critical element in ensuring product quality, stability, and safety throughout the distribution and storage process. The selection and qualification of appropriate packaging materials must consider multiple factors, including product compatibility, stability requirements, and regulatory compliance. Modern pharmaceutical packaging processes incorporate sophisticated technologies and materials designed to protect product integrity while facilitating efficient manufacturing operations.

9.11.2 Primary Packaging Materials

The selection of primary packaging materials requires careful consideration of multiple factors, including material properties, compatibility with pharmaceutical products, and regulatory requirements. Modern pharmaceutical packaging materials incorporate advanced barrier properties, innovative delivery mechanisms, and sophisticated tamper-evident features. The qualification of these materials involves comprehensive testing programs that evaluate physical properties, chemical compatibility, and stability characteristics under various environmental conditions.

9.11.3 Secondary Packaging Considerations

Secondary packaging serves multiple functions in pharmaceutical products, including protection during distribution, facilitation of proper storage conditions, and communication of essential product information. The design and implementation of effective secondary packaging solutions require careful consideration of various factors, including environmental conditions, distribution requirements, and regulatory compliance. Modern secondary packaging solutions incorporate innovative materials and designs that enhance product protection while optimizing distribution efficiency.

9.11.4 Packaging Process Validation

The validation of pharmaceutical packaging processes represents a critical aspect of quality assurance programs. These validation activities must address various aspects of the packaging operation, including equipment qualification, process parameters, and quality control procedures. The development and implementation of comprehensive validation protocols ensure that packaging processes consistently produce products that meet quality requirements while maintaining regulatory compliance.

9.11.5 Quality Control in Packaging Operations

Quality control in pharmaceutical packaging operations encompasses multiple activities designed to ensure consistent product quality and compliance with regulatory requirements. These activities include in-process testing, finished product inspection, and comprehensive documentation of all packaging operations. Modern quality control systems incorporate automated inspection technologies, advanced testing methodologies, and sophisticated data management capabilities to ensure consistent product quality.

9.11.6 Material Testing and Qualification

The testing and qualification of packaging materials represent critical activities in ensuring product quality and regulatory compliance. These activities must address various material characteristics, including physical properties, chemical composition, and barrier performance. The implementation of comprehensive testing programs ensures that packaging materials consistently meet quality requirements while maintaining product stability throughout the intended shelf life (Table 9.4).

9.12 QUALIFICATION OF A SERVO-CONTROLLED BLISTER PACKAGING LINE

9.12.1 Introduction to Servo-Controlled Blister Packaging

The qualification of servo-controlled blister packaging lines represents a critical aspect of pharmaceutical manufacturing, incorporating advanced automation

TABLE 9.4
Testing Parameters, Methods, and Acceptance Criteria for Packaging Material Qualification

Testing Parameter	Test Method	Acceptance Criteria	Frequency
Barrier properties	ASTM F1249	≤0.01 g/package/day	Each lot
Chemical compatibility	USP <661>	No interactions	Validation
Mechanical strength	ASTM D638	>15 MPa	Each lot
Container closure	USP <671>	No leakage	Each lot

technologies with precise control mechanisms to ensure product integrity and packaging quality. Modern servo-controlled blister packaging systems utilize sophisticated motion control algorithms and feedback mechanisms to achieve unprecedented levels of accuracy and consistency in pharmaceutical packaging operations. These systems integrate multiple servo motors and controllers, each precisely coordinated to perform specific functions within the packaging process, from material feeding and forming to sealing and cutting operations.

9.12.2 Design Qualification and System Architecture

The design qualification of servo-controlled blister packaging lines encompasses a comprehensive evaluation of system architecture, component integration, and control mechanisms. The fundamental architecture typically incorporates multiple servo axes, each responsible for specific mechanical operations within the packaging process. These systems utilize advanced programmable logic controller (PLC) platforms integrated with sophisticated servo drive networks, enabling precise synchronization of multiple mechanical operations. The design phase must consider factors such as material specifications, production speed requirements, changeover flexibility, and compliance with current Good Manufacturing Practice (cGMP) guidelines (Figure 9.3).

9.12.3 Installation Qualification and Technical Parameters

The installation qualification process for servo-controlled blister packaging lines involves rigorous verification of mechanical, electrical, and control system installations. This phase requires detailed documentation of all installation parameters, including servo motor specifications, drive configurations, control system settings, and safety system implementations. The technical parameters established during installation qualification serve as the foundation for subsequent operational qualification and performance verification activities. Installation qualification procedures must verify proper implementation of all control interfaces, safety interlocks, and monitoring systems essential for reliable operation.

9.12.4 Operational Qualification and Performance Standards

Operational qualification of servo-controlled blister packaging systems focuses on verifying system performance across the entire operational envelope. This includes comprehensive testing of servo control algorithms, motion profiles, and synchronization parameters under various operating conditions. The qualification process must validate the system's ability to maintain precise control over critical packaging parameters, including forming temperature, sealing pressure, and material feed rates. Performance standards established during operational qualification typically include specifications for positioning accuracy, speed stability, and overall system reliability.

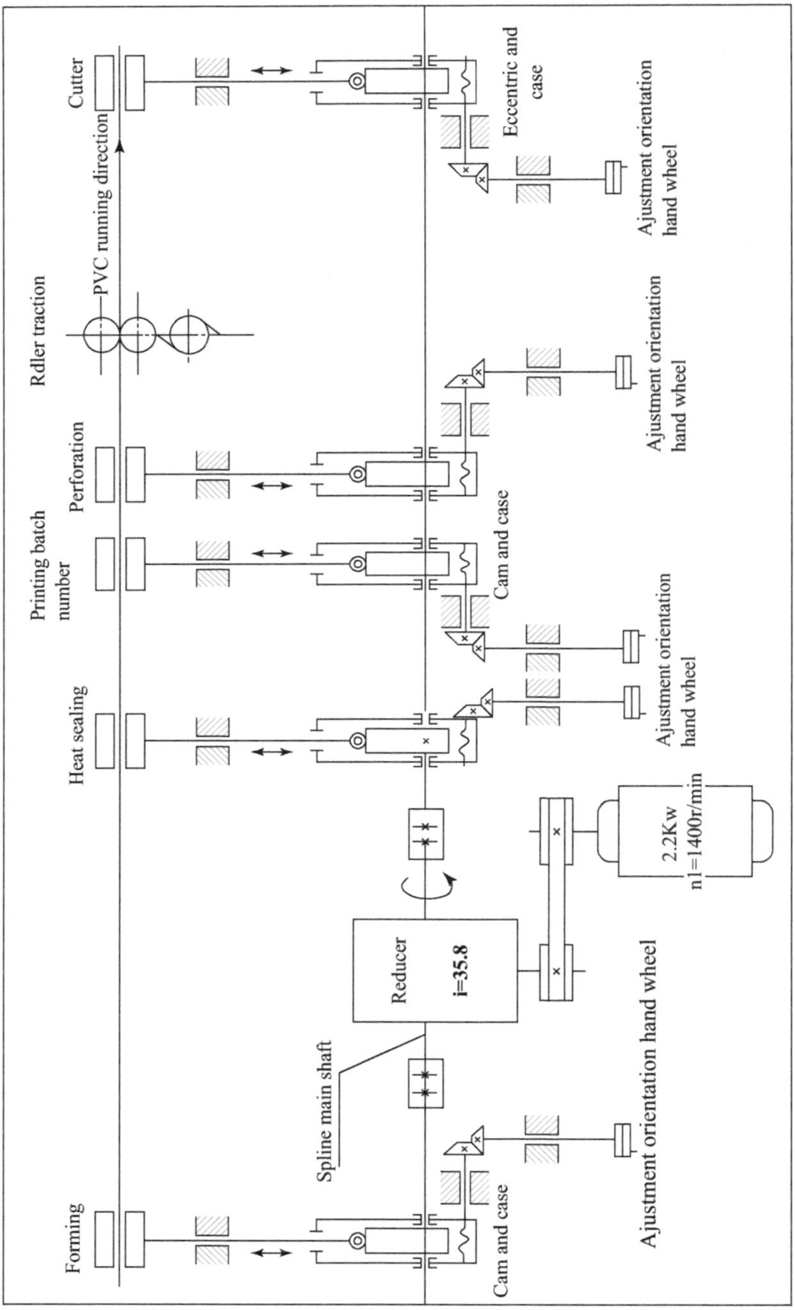

FIGURE 9.3 Architectural overview of servo-controlled blister packaging system with control integration points.

9.12.5 Performance Qualification and Production Validation

The performance qualification phase validates the system's capability to consistently produce quality packaging under actual production conditions. This phase involves extensive testing across multiple production runs, incorporating various product formats and packaging materials. Performance qualification must demonstrate consistent achievement of critical quality attributes, including seal integrity, product protection, and packaging appearance. The validation process typically includes challenge testing to verify system performance under worst-case scenarios and boundary conditions.

9.13 BLOW-FILL-SEAL (BFS) TECHNOLOGY

9.13.1 Fundamentals of BFS Technology

Blow-fill-seal technology represents a sophisticated aseptic packaging solution that integrates container formation, filling, and sealing into a continuous, automated process. This advanced manufacturing technology offers significant advantages in terms of product protection, contamination control, and operational efficiency. The BFS process begins with the extrusion of sterile plastic material, typically low-density polyethylene (LDPE) or polypropylene (PP), which is then formed into containers through a precise combination of mechanical and pneumatic operations.

9.13.2 Process Control and Quality Parameters

The control of BFS operations requires sophisticated monitoring and adjustment of multiple process parameters, including extrusion temperature, forming pressure, fill accuracy, and sealing conditions. Quality parameters must be continuously monitored to ensure consistent container formation, accurate filling, and reliable sealing. The process control system must maintain precise coordination between mechanical operations, filling systems, and sealing mechanisms to achieve optimal product quality and packaging integrity.

9.13.3 Aseptic Processing and Environmental Controls

BFS technology incorporates advanced aseptic processing capabilities through the integration of sophisticated environmental control systems and sterile air handling mechanisms. The aseptic processing zone within BFS systems requires carefully controlled air quality, temperature, and pressure differentials to maintain sterile conditions throughout the packaging operation. Environmental monitoring systems play a crucial role in verifying and documenting the maintenance of aseptic conditions during production.

9.13.4 Material Considerations and Container Design

The selection of appropriate materials and container designs for BFS applications requires careful consideration of multiple factors, including product compatibility,

TABLE 9.5
BFS Material Selection Criteria and Performance Requirements

Material Property	Requirement	Test Method	Acceptance Criteria
Clarity	Clear	Visual	No visible defects
Barrier properties	High	ASTM F1249	<0.10 g/m²/day
Heat stability	Stable	DSC	Tm > 160°C
Chemical resistance	Inert	USP <661>	No interaction

stability requirements, and sterilization methods. Container design must accommodate both functional requirements and manufacturing constraints while ensuring product protection and user convenience. Material specifications must address factors such as clarity, barrier properties, and stability under various environmental conditions (Table 9.5).

9.13.5 Validation and Quality Assurance

The validation of BFS processes encompasses comprehensive testing and verification of all critical process parameters and quality attributes. Validation protocols must address aspects such as container integrity, fill accuracy, and sterility assurance. Quality assurance programs for BFS operations typically include regular monitoring of environmental conditions, process parameters, and finished product quality. Statistical process control methods are employed to track and analyze key performance indicators and maintain consistent product quality.

9.14 ARTIFICIAL INTELLIGENCE IN PRODUCTION AND PACKAGING

9.14.1 Introduction to AI Integration in Pharmaceutical Manufacturing

The integration of artificial intelligence (AI) in pharmaceutical production and packaging represents a transformative advancement in manufacturing technology, fundamentally reshaping traditional approaches to quality control, process optimization, and regulatory compliance. Modern pharmaceutical facilities are increasingly adopting AI-driven solutions to enhance operational efficiency, reduce human error, and maintain consistent product quality throughout the manufacturing process. This technological revolution extends beyond simple automation, encompassing sophisticated machine learning algorithms, computer vision systems, and predictive analytics that work in concert to create an intelligent manufacturing ecosystem. The implementation of AI systems in pharmaceutical manufacturing has demonstrated significant improvements in production efficiency, quality assurance, and regulatory compliance while simultaneously reducing operational costs and minimizing product waste.

9.14.2 Real-ime Monitoring Systems and Process Control

The deployment of AI-powered real-time monitoring systems represents a fundamental shift in how pharmaceutical manufacturers approach process control and quality assurance. These sophisticated systems utilize an array of sensors and data collection points throughout the production line, continuously gathering information about critical process parameters, environmental conditions, and product characteristics. Advanced neural networks process this data stream in real time, identifying patterns and trends that might be imperceptible to human operators. The integration of machine learning algorithms enables these systems to adapt and improve their performance over time, learning from historical data to predict potential quality issues before they manifest in the final product.

9.14.3 Computer Vision and Quality Inspection

The application of AI-driven computer vision systems has revolutionized quality inspection processes in pharmaceutical packaging. These sophisticated systems employ deep learning algorithms to analyze high-resolution images and video feeds of products moving through the packaging line, detecting defects and irregularities with unprecedented accuracy and speed. The implementation of computer vision technology extends beyond simple visual inspection, incorporating multispectral imaging and 3D scanning capabilities to evaluate product integrity from multiple perspectives. These systems can identify subtle variations in package sealing, label placement, and product appearance, ensuring consistent quality while maintaining high production speeds.

9.14.4 Predictive Maintenance and Equipment Optimization

Artificial intelligence has transformed traditional maintenance approaches through the implementation of predictive maintenance systems. These sophisticated platforms analyze equipment performance data, vibration patterns, and operational parameters to forecast potential failures before they occur. The integration of machine learning algorithms enables these systems to identify subtle changes in equipment behavior that might indicate impending issues, allowing maintenance teams to address problems proactively rather than reactively. This predictive approach has significantly reduced unplanned downtime, extended equipment lifespan, and optimized maintenance scheduling across pharmaceutical manufacturing facilities.

9.14.5 Data Analytics and Process Optimization

The implementation of AI-powered data analytics platforms has enabled pharmaceutical manufacturers to optimize their production processes with unprecedented precision. These systems analyze vast quantities of historical and real-time data to identify opportunities for process improvement, waste reduction, and efficiency enhancement. Advanced machine learning algorithms process information from multiple sources, including production equipment, environmental monitoring systems, and quality control data, to create comprehensive models of manufacturing operations. These models enable manufacturers to optimize process parameters, reduce energy consumption, and minimize material waste while maintaining product quality and regulatory compliance.

9.14.6 Regulatory Compliance and Documentation

Artificial intelligence systems play an increasingly crucial role in ensuring regulatory compliance and maintaining accurate documentation throughout the manufacturing process. These platforms automate the collection and analysis of compliance-related data, generating detailed reports and maintaining electronic batch records with minimal human intervention. The implementation of natural language processing algorithms enables these systems to interpret regulatory requirements and automatically flag potential compliance issues before they become significant problems. This automated approach to compliance management has significantly reduced the administrative burden on pharmaceutical manufacturers while improving the accuracy and completeness of regulatory documentation.

9.15 CONCLUSION

The integration of artificial intelligence in pharmaceutical production and packaging represents a significant advancement in manufacturing technology, fundamentally transforming how pharmaceutical products are manufactured, inspected, and packaged. The implementation of AI-driven systems has enabled manufacturers to achieve unprecedented levels of quality control, process optimization, and regulatory compliance while reducing operational costs and minimizing waste. The continued evolution of AI technology promises further improvements in manufacturing efficiency and product quality, driving the pharmaceutical industry toward an increasingly automated and intelligent future.

The comprehensive integration of artificial intelligence across all aspects of pharmaceutical manufacturing has created a new paradigm in production efficiency and quality assurance. From real-time monitoring and process control to predictive maintenance and regulatory compliance, AI systems have demonstrated their ability to enhance manufacturing operations while reducing costs and improving product quality. As these technologies continue to evolve, their impact on pharmaceutical manufacturing will only grow, driving further innovations in production methods and quality control processes.

10 GMP in Regulation

10.1 INTRODUCTION

The pharmaceutical industry operates within one of the most stringently regulated environments in the modern manufacturing sector, with Good Manufacturing Practice (GMP) serving as the fundamental framework ensuring product quality, safety, and efficacy. The regulatory landscape has undergone substantial transformation since the inception of formal pharmaceutical regulations in the early 20th century, evolving from simple quality control measures to comprehensive systems encompassing every aspect of pharmaceutical production. This evolution reflects the industry's growing complexity, technological advancement, and the increasing globalization of pharmaceutical manufacturing and distribution networks. The modern regulatory framework represents a sophisticated interplay between national authorities, international organizations, and industry stakeholders, all working in concert to ensure the highest standards of pharmaceutical quality.

The regulatory landscape encompasses multiple layers of oversight, each contributing to a comprehensive structure of quality assurance and control. National regulatory authorities serve as the primary enforcers of pharmaceutical quality standards within their jurisdictions, while international organizations work to harmonize these requirements across borders. This multilayered approach has become increasingly important as pharmaceutical supply chains extend globally, necessitating coordinated regulatory efforts to ensure consistent quality standards across different regions and manufacturing locations.

The modern GMP framework extends beyond traditional quality control measures, incorporating risk management principles, quality by design approaches, and advanced technological solutions. This comprehensive approach ensures that quality is built into products from the earliest stages of development through to final distribution. The integration of modern technologies, particularly artificial intelligence and advanced data analytics, has introduced new possibilities for regulatory compliance while simultaneously creating new challenges that regulatory frameworks must address (Figure 10.1).

The significance of GMP regulations in ensuring public health cannot be overstated. These regulations provide the foundation for pharmaceutical quality assurance, ensuring that medicines consistently meet the required quality standards. The regulatory framework must balance the need for stringent quality control with the industry's ability to innovate and efficiently produce medicines. This balance becomes particularly crucial as new manufacturing technologies emerge and production processes become increasingly complex.

DOI: 10.1201/9781003624592-10

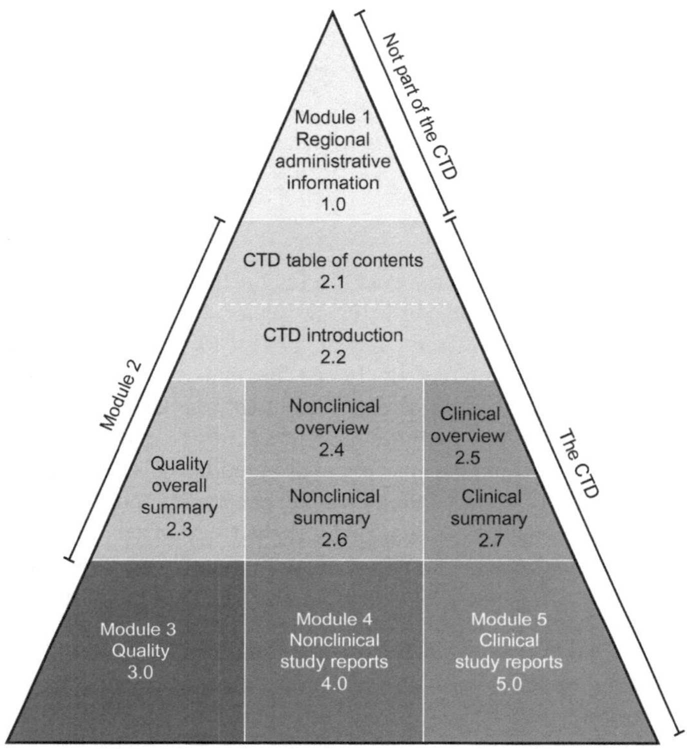

FIGURE 10.1 Hierarchical structure of global pharmaceutical regulation.

Recent developments in pharmaceutical manufacturing, including the advent of personalized medicine, continuous manufacturing processes, and advanced therapy medicinal products, have necessitated ongoing evolution in regulatory approaches. The traditional GMP framework has demonstrated remarkable adaptability, incorporating new requirements while maintaining its fundamental focus on quality, safety, and efficacy. This adaptability ensures that regulatory oversight remains relevant and effective in the face of technological advancement and changing manufacturing paradigms.

The interconnected nature of modern pharmaceutical manufacturing has led to increased emphasis on international cooperation and harmonization of regulatory requirements. This cooperation manifests through various initiatives, including the International Council for Harmonisation of Technical Requirements for Pharmaceuticals for Human Use (ICH), the Pharmaceutical Inspection Co-operation Scheme (PIC/S), and numerous bilateral agreements between regulatory authorities. These collaborative efforts aim to reduce regulatory burden while maintaining high quality standards across global markets.

10.2 INFORMATION ON NATIONAL BODIES AND PHARMACEUTICAL ASSOCIATIONS

The global pharmaceutical regulatory landscape is characterized by a complex network of national regulatory bodies and pharmaceutical associations that work in concert to ensure the safety, efficacy, and quality of pharmaceutical products. These organizations serve as the primary architects and enforcers of pharmaceutical regulations within their respective jurisdictions while simultaneously participating in international harmonization efforts. The role of national regulatory authorities has evolved significantly over the past decades, transitioning from simple enforcement agencies to sophisticated organizations that employ risk-based approaches and modern technological solutions in their regulatory oversight.

National regulatory bodies, such as the Food and Drug Administration (FDA) in the United States, the European Medicines Agency (EMA) in the European Union, and the Pharmaceuticals and Medical Devices Agency (PMDA) in Japan, operate under distinct legislative frameworks that reflect their specific national priorities and regulatory philosophies. These organizations have developed comprehensive systems for evaluating pharmaceutical manufacturing facilities, reviewing product applications, and monitoring post-market safety. Their approaches to regulatory oversight have become increasingly sophisticated, incorporating risk-based methodologies and advanced analytical tools to optimize their inspection and enforcement activities (Figure 10.2).

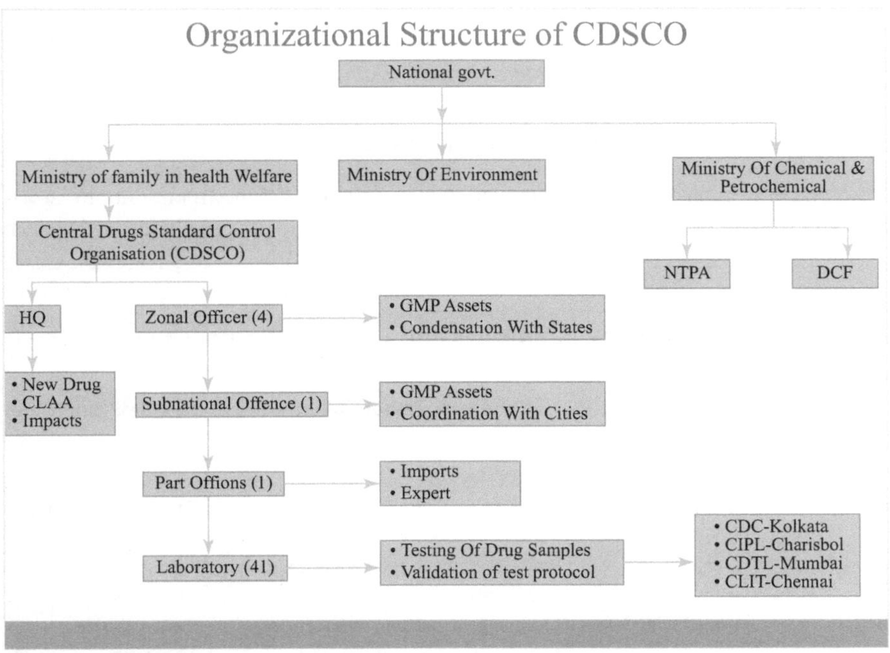

FIGURE 10.2 Organizational structure of major national regulatory authorities.

The collaboration between national regulatory authorities has become increasingly important in the context of globalized pharmaceutical manufacturing. This collaboration manifests through various mechanisms, including mutual recognition agreements (MRAs), confidentiality agreements, and joint inspection programs. These collaborative arrangements help reduce regulatory burden while ensuring consistent application of GMP standards across different jurisdictions. The sharing of inspection reports, quality defect information, and regulatory intelligence has become standard practice among major regulatory authorities, facilitating more efficient oversight of global pharmaceutical supply chains.

Pharmaceutical associations play a crucial complementary role to regulatory authorities, serving as bridges between industry and regulators while providing valuable technical expertise and industry perspectives in regulatory development. Organizations such as the International Society for Pharmaceutical Engineering (ISPE), the Parenteral Drug Association (PDA), and various national pharmaceutical manufacturers' associations contribute significantly to the development of technical guidelines and best practices. These associations often work closely with regulatory authorities to provide industry feedback on proposed regulations and help develop practical implementation strategies for new regulatory requirements.

The interaction between national bodies and international organizations has led to the development of sophisticated regulatory networks that facilitate information sharing and collaborative enforcement actions. These networks have become increasingly important as pharmaceutical supply chains have grown more complex and globally distributed. The ability to coordinate regulatory actions across jurisdictions has become crucial for addressing quality issues that may affect multiple markets simultaneously (Table 10.1).

10.3 OVERVIEW OF PHARMACOPOEIA

The concept of pharmacopoeia represents one of the most fundamental and historically significant aspects of pharmaceutical regulation and standardization. Modern

TABLE 10.1
Major National Regulatory Authorities and Their Key Responsibilities

Regulatory Authority	Jurisdiction	Primary Responsibilities	Key GMP Guidelines
FDA	United States	Drug approval, facility inspection, enforcement	21 CFR Parts 210/211
EMA	European Union	Scientific evaluation, supervision, safety monitoring	EU GMP Guidelines
PMDA	Japan	Product review, safety monitoring, GMP inspection	Japanese GMP Ordinance
MHRA	United Kingdom	Licensing, inspection, enforcement	Orange Guide
Health Canada	Canada	Drug review, facility inspection, compliance monitoring	GUI-0001

pharmacopoeias serve as comprehensive compendia of drug standards, encompassing detailed specifications for pharmaceutical substances, formulated preparations, and analytical methodologies. These authoritative publications play a crucial role in establishing and maintaining quality standards across the global pharmaceutical industry, serving as essential references for manufacturers, regulators, and healthcare professionals alike.

The development and evolution of pharmacopoeial standards reflect the advancing complexity of pharmaceutical science and manufacturing technology. Each pharmacopoeia maintains extensive monographs detailing specific requirements for individual substances, including physical and chemical properties, identification methods, purity criteria, and analytical procedures. These standards are continuously updated to incorporate new scientific knowledge, analytical technologies, and quality requirements, ensuring their relevance in modern pharmaceutical manufacturing (Figure 10.3).

The international landscape of pharmacopoeial authorities presents a complex network of cooperation and harmonization efforts. The major pharmacopoeias, including the United States Pharmacopeia (USP), European Pharmacopoeia (Ph. Eur.), and Japanese Pharmacopoeia (JP), maintain extensive collaboration through the Pharmacopeial Discussion Group (PDG) and other international forums. This collaboration aims to reduce redundancy in testing requirements and establish globally accepted standards, facilitating international trade while maintaining high-quality standards.

Modern pharmacopoeial requirements extend beyond traditional chemical and physical testing parameters to encompass biological standards, impurity profiles, and specialized requirements for advanced therapeutic products. This expansion reflects the growing diversity of pharmaceutical products and the increasing sophistication of analytical methodologies. The integration of modern analytical techniques, including high-performance liquid chromatography (HPLC), mass spectrometry,

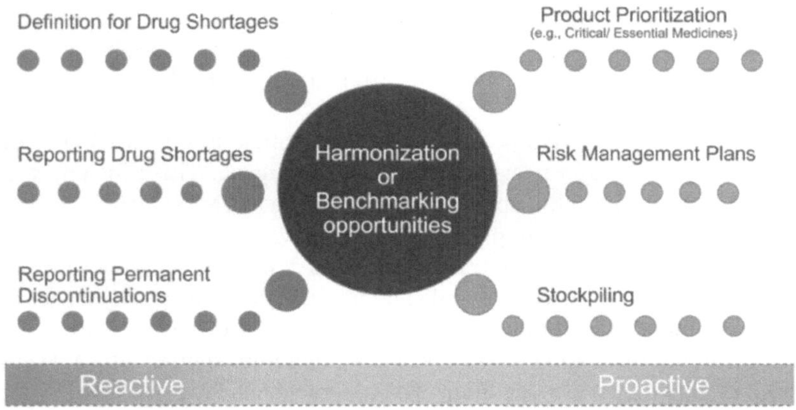

FIGURE 10.3 Global pharmacopoeial integration and harmonization network.

and spectroscopic methods, has significantly enhanced the precision and specificity of pharmacopoeial testing requirements.

The harmonization of pharmacopoeial standards has become increasingly important in the context of globalized pharmaceutical manufacturing. Differences in pharmacopoeial requirements between major markets can create significant challenges for manufacturers, necessitating multiple testing protocols and validation procedures. The ongoing efforts to harmonize these requirements through international collaboration represent a crucial development in reducing regulatory burden while maintaining rigorous quality standards.

10.4 EU DIRECTIVES AND GUIDELINES

The European Union's regulatory framework for pharmaceutical manufacturing represents one of the most comprehensive and influential systems globally. The EU GMP guidelines, enshrined in EudraLex Volume 4, provide detailed requirements for pharmaceutical manufacturing, covering all aspects from facility design to quality management systems. These guidelines are continuously updated to reflect technological advancements and emerging quality concerns, serving as a model for regulatory frameworks worldwide.

The structure of EU pharmaceutical regulation operates on multiple levels, incorporating directives, regulations, and guidelines that form a cohesive regulatory framework. Directive 2001/83/EC and its amendments establish the fundamental requirements for medicinal products in the EU, while specific GMP requirements are detailed in Commission Directive 2003/94/EC. These legal instruments are supported by extensive guidance documents, including annexes addressing specific aspects of pharmaceutical manufacturing and control.

The EU's risk-based approach to pharmaceutical regulation has significantly influenced global regulatory practices. This approach emphasizes the importance of understanding and controlling critical process parameters, implementing appropriate quality control measures, and maintaining comprehensive documentation systems. The concept of quality risk management (QRM), as outlined in ICH Q9 and incorporated into EU GMP guidelines, provides a systematic framework for identifying, assessing, and controlling quality risks throughout the manufacturing process.

Recent developments in EU pharmaceutical regulation have focused on addressing emerging challenges in pharmaceutical manufacturing. These include specific requirements for advanced therapy medicinal products (ATMPs), continuous manufacturing processes, and the implementation of new analytical technologies. The adaptability of the EU regulatory framework ensures its continued relevance in the face of technological advancement while maintaining stringent quality standards.

The EU's mutual recognition agreements (MRAs) with major pharmaceutical markets, including the United States and Japan, represent significant achievements in international regulatory cooperation. These agreements facilitate trade by reducing duplicate inspections and recognizing the equivalence of GMP compliance systems between participating authorities. The expansion and maintenance of these agreements demonstrate the EU's commitment to international regulatory harmonization while maintaining high quality standards.

The implementation of EU GMP requirements necessitates a comprehensive quality management system that integrates all aspects of pharmaceutical manufacturing. This includes detailed requirements for personnel training, facility design, equipment qualification, process validation, and quality control testing. The emphasis on systematic documentation and traceability ensures that manufacturing processes remain consistent and compliant with regulatory requirements.

10.5 UNITED STATES: CFR AND FDA GUIDELINES

The US regulatory framework for pharmaceutical manufacturing, primarily governed by the Food and Drug Administration (FDA), represents one of the most comprehensive and influential systems globally. The Code of Federal Regulations (CFR), particularly Title 21, provides the legal foundation for pharmaceutical manufacturing requirements in the United States. Parts 210 and 211 of Title 21 specifically outline the current Good Manufacturing Practice (cGMP) regulations that set the standard for pharmaceutical production. These regulations have evolved significantly since their initial implementation, incorporating new scientific understanding, technological advances, and lessons learned from industry experience.

10.5.1 Structure and Implementation of FDA Guidelines

The FDA's approach to GMP regulation emphasizes a risk-based methodology that focuses on critical quality attributes and process parameters. The guidelines cover every aspect of pharmaceutical manufacturing, from facility design and equipment qualification to process validation and quality control testing. The FDA has developed a sophisticated system of guidance documents that complement the CFR requirements, providing detailed interpretations and recommendations for implementing GMP regulations. These guidance documents, while not legally binding, represent the FDA's current thinking on various aspects of pharmaceutical manufacturing and serve as valuable resources for industry compliance (Figure 10.4).

The FDA's inspection program serves as the primary mechanism for evaluating compliance with GMP requirements. The risk-based inspection model prioritizes facilities based on various factors, including product complexity, manufacturing process sophistication, and compliance history. This approach ensures efficient allocation of regulatory resources while maintaining comprehensive oversight of the pharmaceutical manufacturing sector. The FDA's Form 483 observations and Warning Letters provide insights into common compliance issues and agency expectations, serving as valuable learning tools for the entire industry.

10.5.2 Quality Systems Approach

The FDA's quality systems approach integrates various aspects of pharmaceutical manufacturing under a comprehensive quality management framework. This

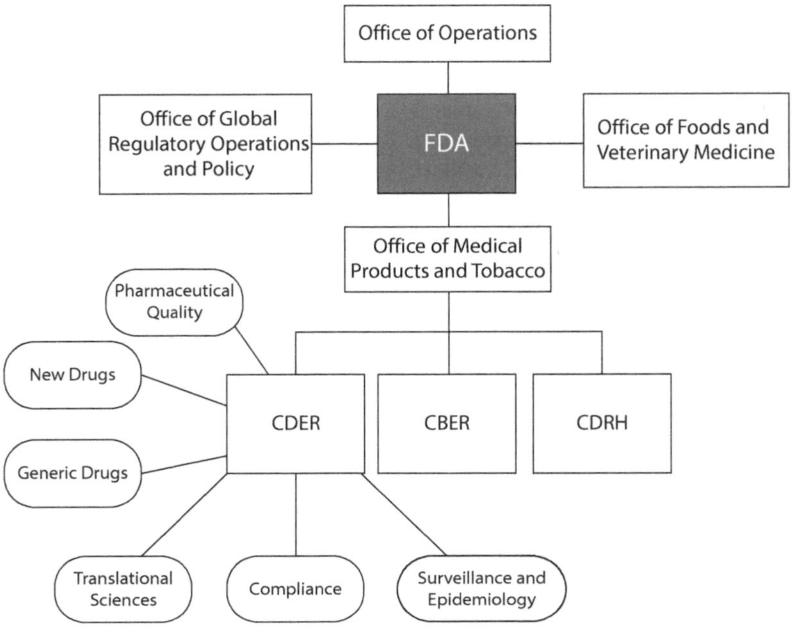

FIGURE 10.4 Hierarchical structure of FDA regulatory framework and implementation process.

approach emphasizes the importance of establishing robust quality systems that encompass all aspects of production, from raw material testing to finished product release. The quality systems model promotes a holistic view of pharmaceutical manufacturing, recognizing the interconnected nature of various manufacturing operations and their impact on product quality.

10.5.3 Process Validation and Control Strategies

Process validation requirements under FDA guidelines have evolved to reflect current understanding of pharmaceutical manufacturing science. The life cycle approach to process validation, as outlined in the FDA's guidance documents, emphasizes ongoing verification of process performance and product quality. This approach recognizes that process validation is not a one-time event but a continuous activity that spans the entire product life cycle.

10.6 ICH Guidelines

The International Council for Harmonisation of Technical Requirements for Pharmaceuticals for Human Use (ICH) represents a unique partnership between regulatory authorities and pharmaceutical industry representatives. ICH guidelines have become the global standard for pharmaceutical development, manufacturing, and quality control, providing a harmonized approach to pharmaceutical regulation across major markets.

10.6.1 Quality Guidelines (Q Series)

The ICH Q series guidelines form the backbone of modern pharmaceutical quality systems, providing a comprehensive framework for development, manufacturing, and control strategies. ICH Q8 (pharmaceutical development), Q9 (quality risk management), and Q10 (pharmaceutical quality system) constitute the primary pillars of this framework, promoting a systematic approach to pharmaceutical quality management.

The implementation of these guidelines has fundamentally transformed how pharmaceutical companies approach product development and manufacturing. The quality by design (QbD) concepts outlined in ICH Q8 encourage a systematic approach to product and process understanding, emphasizing the importance of identifying critical quality attributes and process parameters during development. This scientific, risk-based approach helps ensure consistent product quality while providing flexibility in manufacturing operations.

10.6.2 Harmonization and Implementation

The harmonization process through ICH has significantly reduced regulatory barriers while maintaining high quality standards. The guidelines provide a common technical language and shared expectations across different regulatory jurisdictions, facilitating global pharmaceutical development and manufacturing operations. The implementation of ICH guidelines requires significant organizational commitment and resources, but the benefits of harmonized approaches often justify these investments.

10.6.3 Risk Management and Quality Systems

ICH Q9 provides a comprehensive framework for quality risk management in pharmaceutical manufacturing. This systematic approach to identifying, evaluating, and controlling potential quality risks has become an integral part of pharmaceutical quality systems. The integration of risk management principles throughout the product life cycle helps organizations allocate resources effectively while maintaining focus on critical quality aspects.

10.6.4 Continuous Improvement and Knowledge Management

The ICH guidelines emphasize the importance of continuous improvement and knowledge management in pharmaceutical manufacturing. The concept of the pharmaceutical quality system, as outlined in ICH Q10, provides a framework for implementing modern quality management principles in pharmaceutical operations. This framework promotes the use of knowledge gained throughout the product life cycle to drive continuous improvement in manufacturing processes and control strategies.

10.7 WHO GUIDELINES

The World Health Organization (WHO) plays a pivotal role in establishing and promoting global standards for pharmaceutical manufacturing through its comprehensive Good Manufacturing Practice (GMP) guidelines. These guidelines serve as the foundation for many national regulatory frameworks, particularly in developing nations and emerging markets. The WHO's approach to pharmaceutical regulation emphasizes accessibility and adaptability while maintaining rigorous quality standards, making it particularly relevant for diverse manufacturing environments worldwide.

The WHO GMP guidelines provide a comprehensive framework that addresses all aspects of pharmaceutical manufacturing, from facility design to quality management systems. These guidelines are particularly significant because they consider the unique challenges faced by manufacturers in different regions, including varying technological capabilities and resource constraints. The WHO's approach balances the need for stringent quality control with practical implementation considerations, making the guidelines applicable across diverse manufacturing settings.

The WHO's role in promoting international harmonization of pharmaceutical regulations extends beyond merely publishing guidelines. Through its prequalification program, the WHO actively evaluates pharmaceutical products and manufacturers, ensuring they meet international standards for quality, safety, and efficacy. This program has become particularly important for ensuring access to quality medicines in regions where regulatory oversight may be limited or developing.

The WHO GMP guidelines incorporate modern quality management principles while maintaining flexibility in implementation approaches. The guidelines emphasize risk-based decision-making, encouraging manufacturers to develop quality systems appropriate to their specific operations while meeting fundamental quality requirements. This adaptable approach has proven particularly valuable in supporting pharmaceutical manufacturing development in emerging markets while maintaining consistent quality standards.

The organization's commitment to continuous improvement in pharmaceutical regulation is evident in its regular updates and supplementary guidance documents. These updates reflect technological advancements, emerging quality concerns, and lessons learned from implementation experiences worldwide. The WHO's collaborative approach, involving experts from various regions and regulatory authorities, ensures that its guidelines remain relevant and practical while promoting global health objectives.

The WHO's influence extends to capacity building and training initiatives, supporting the development of regulatory expertise in different regions. These efforts help ensure effective implementation of GMP requirements and strengthen global pharmaceutical quality standards. The organization's technical assistance programs have been instrumental in developing robust regulatory systems in many countries.

10.8 ARTIFICIAL INTELLIGENCE IN REGULATORY COMPLIANCE

The integration of artificial intelligence (AI) into pharmaceutical regulatory compliance represents a transformative development in quality assurance and control. AI technologies are increasingly being applied to various aspects of regulatory compliance, from automated document review to predictive maintenance and real-time process monitoring. This technological evolution is reshaping traditional approaches to GMP compliance while introducing new possibilities for enhancing quality assurance systems.

Advanced machine learning algorithms have demonstrated remarkable capabilities in analyzing complex manufacturing data and identifying potential quality issues before they manifest in product defects. These systems can process vast amounts of production data in real time, detecting subtle patterns and correlations that might escape traditional quality control methods. The implementation of AI-driven monitoring systems has significantly enhanced the ability to maintain consistent product quality while optimizing manufacturing processes (Table 10.2).

The application of AI in regulatory document management has revolutionized compliance documentation processes. Natural language processing (NLP) technologies can analyze regulatory submissions, standard operating procedures, and quality control documentation with unprecedented speed and accuracy. These systems help ensure consistency in documentation while identifying potential compliance gaps or areas requiring attention.

The implementation of AI-driven predictive maintenance systems represents another significant advancement in GMP compliance. These systems analyze equipment performance data to predict potential failures or maintenance requirements before they impact product quality. This proactive approach to equipment maintenance has significantly reduced production disruptions while ensuring consistent compliance with GMP requirements.

AI technologies have also transformed batch release processes through the implementation of intelligent review systems. These systems can analyze complex batch records and associated quality control data, identifying potential issues and ensuring compliance with established specifications. The ability to process and analyze large volumes of manufacturing data has significantly improved the efficiency and reliability of batch release decisions.

TABLE 10.2
AI Applications in Pharmaceutical Quality Management

Application Area	AI Technology	Primary Benefits
Document management	Natural language processing	Automated review and verification
Process control	Machine learning	Real-time quality prediction
Risk assessment	Deep learning	Enhanced risk identification
Compliance monitoring	Computer vision	Automated inspection systems
Training management	Expert systems	Personalized learning paths

The integration of AI in environmental monitoring systems has enhanced the ability to maintain clean room conditions and detect potential contamination risks. Advanced sensor networks combined with AI analytics provide real-time monitoring of environmental parameters, allowing for immediate response to any deviations from specified conditions. This technology has particularly benefited sterile manufacturing operations, where maintaining environmental control is crucial for product quality.

The future development of AI applications in regulatory compliance promises even greater possibilities for enhancing pharmaceutical quality assurance. Emerging technologies, including quantum computing and advanced neural networks, may provide new capabilities for process optimization and quality control. However, these developments also raise important considerations regarding validation, data integrity, and regulatory acceptance of AI-driven decision-making systems.

10.9 ARTIFICIAL INTELLIGENCE IN REGULATORY COMPLIANCE

In the rapidly evolving landscape of pharmaceutical manufacturing and regulatory compliance, artificial intelligence has emerged as a transformative force, revolutionizing how organizations approach and maintain regulatory adherence. The integration of AI technologies into regulatory compliance frameworks represents a significant paradigm shift from traditional manual oversight methods to sophisticated, automated systems capable of real-time monitoring, prediction, and adaptation. This fundamental transformation has implications across the entire pharmaceutical manufacturing ecosystem, from documentation management to process control and quality assurance.

10.9.1 AI-Driven Documentation Management and Control

The management and control of regulatory documentation represent one of the most challenging aspects of GMP compliance. Artificial intelligence systems have revolutionized this domain by introducing sophisticated natural language processing capabilities that can automatically analyze, categorize, and validate regulatory documents. These systems employ advanced algorithms to scan through vast amounts of documentation, identifying potential inconsistencies, gaps in compliance, and areas requiring attention. The implementation of AI-driven documentation management extends beyond simple document storage and retrieval, encompassing intelligent version control, automated updates based on regulatory changes, and predictive analytics for document life cycle management.

Modern AI systems can process multiple document formats, including standard operating procedures, batch records, validation protocols, and regulatory submissions, maintaining their integrity while ensuring compliance with current regulatory requirements. These systems learn from historical data and patterns, continuously improving their ability to identify potential compliance issues before they manifest as actual violations. The integration of machine learning algorithms enables these systems to adapt to new regulatory requirements and organizational changes, ensuring sustained compliance in an ever-evolving regulatory landscape (Figure 10.5).

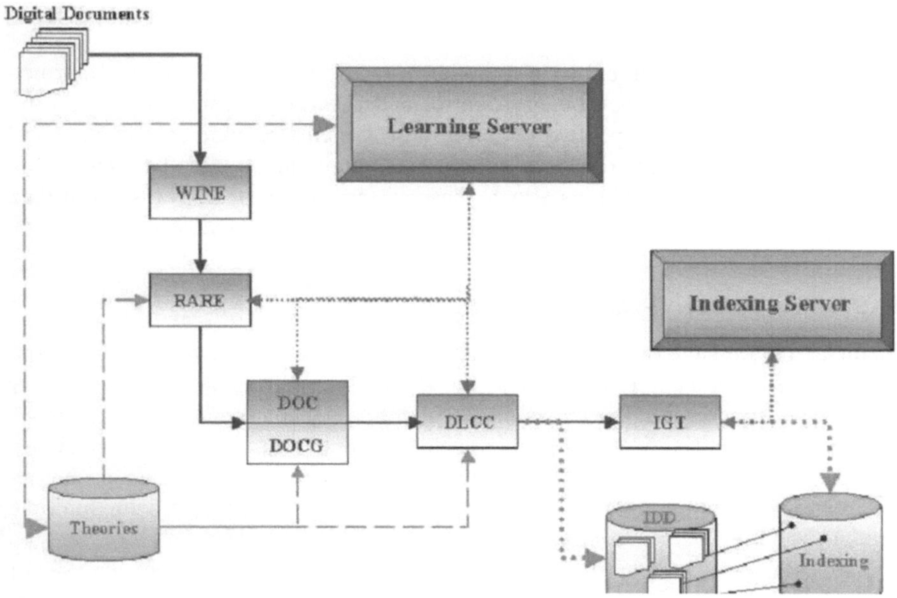

FIGURE 10.5 AI-powered documentation management system architecture.

10.9.2 Predictive Compliance Monitoring

Artificial intelligence has enabled the development of sophisticated predictive compliance monitoring systems that can anticipate potential regulatory issues before they occur. These systems analyze historical compliance data, current operational parameters, and regulatory requirements to identify patterns and trends that might indicate emerging compliance risks. The predictive capabilities extend to various aspects of pharmaceutical manufacturing, including equipment maintenance, process parameters, and quality control measures.

The integration of predictive compliance monitoring with manufacturing execution systems (MES) and quality management systems (QMS) creates a comprehensive oversight mechanism that can detect subtle deviations from established parameters. This early warning system allows organizations to take proactive measures to address potential compliance issues, reducing the risk of regulatory violations and ensuring continuous compliance with GMP requirements.

10.9.3 Real-Time Compliance Verification

Real-time compliance verification represents one of the most significant applications of artificial intelligence in regulatory compliance. Advanced AI systems can monitor manufacturing processes, environmental conditions, and quality parameters in real time, comparing them against regulatory requirements and internal specifications. These systems can instantly identify deviations from established parameters and initiate appropriate corrective actions, ensuring continuous compliance with regulatory requirements.

The implementation of real-time compliance verification systems has transformed how organizations approach quality control and assurance. Instead of relying on periodic audits and retrospective reviews, these systems provide continuous oversight, enabling immediate response to potential compliance issues. The integration of machine learning algorithms allows these systems to learn from historical data and improve their ability to detect and predict compliance-related issues.

10.9.4 Automated Regulatory Intelligence

Artificial intelligence has revolutionized how organizations gather, analyze, and implement regulatory intelligence. Advanced AI systems can continuously monitor regulatory databases, guidance documents, and industry communications, automatically identifying relevant changes and updates that may impact organizational compliance. These systems can analyze the implications of regulatory changes across different jurisdictions, helping organizations maintain compliance with multiple regulatory frameworks simultaneously.

The automated regulatory intelligence capabilities extend to the analysis of enforcement actions, warning letters, and regulatory trends, providing organizations with valuable insights into evolving regulatory expectations and focus areas. This proactive approach to regulatory intelligence enables organizations to adapt their compliance strategies effectively, reducing the risk of regulatory violations and ensuring sustained compliance with current requirements.

10.9.5 Risk Assessment and Management

AI-driven risk assessment and management systems have transformed how organizations identify, evaluate, and mitigate compliance risks. These systems employ sophisticated algorithms to analyze historical data, current operations, and regulatory requirements, generating comprehensive risk assessments that consider multiple factors and their interrelationships. The integration of machine learning capabilities enables these systems to learn from historical risk patterns and outcomes, continuously improving their ability to identify and predict potential compliance risks.

10.9.6 Compliance Training and Education

Artificial intelligence has revolutionized how organizations approach compliance training and education. Advanced AI systems can create personalized training programs based on individual roles, responsibilities, and learning patterns. These systems can adapt training content based on regulatory changes, organizational requirements, and individual performance, ensuring that personnel maintain current knowledge of regulatory requirements and compliance procedures.

The implementation of AI-driven training systems extends beyond traditional learning management systems, incorporating interactive elements, scenario-based learning, and real-time assessment capabilities. These systems can track individual progress, identify knowledge gaps, and provide targeted interventions to ensure comprehensive understanding of regulatory requirements and compliance procedures.

10.10 CONCLUSION

The integration of Good Manufacturing Practice (GMP) regulations within the pharmaceutical industry represents a complex and evolving landscape that continues to shape the future of pharmaceutical manufacturing and quality assurance. This chapter has provided a comprehensive examination of the various aspects of GMP regulations, from national bodies to emerging technological solutions, highlighting the critical role these frameworks play in ensuring product quality and patient safety.

The regulatory landscape has demonstrated remarkable adaptability in response to technological advancements and changing manufacturing paradigms. The evolution from traditional quality control approaches to sophisticated, risk-based systems reflects the industry's commitment to maintaining the highest standards of pharmaceutical quality while embracing innovation. The harmonization efforts across different regions, while respecting jurisdiction-specific requirements, have contributed to a more cohesive global regulatory framework.

The emergence of artificial intelligence and advanced technological solutions has marked a significant transformation in regulatory compliance approaches. These innovations have not only enhanced the efficiency and effectiveness of compliance processes but have also introduced new possibilities for proactive risk management and real-time monitoring. The integration of AI-driven systems in documentation management, predictive compliance monitoring, and automated regulatory intelligence has revolutionized how organizations approach and maintain regulatory compliance.

The role of international organizations and collaborative initiatives in shaping the global regulatory landscape cannot be understated. The efforts of organizations such as ICH, PIC/S, and WHO have been instrumental in promoting consistency and standardization across different regions while maintaining focus on patient safety and product quality. These collaborative efforts have facilitated the development of harmonized guidelines that address the challenges of modern pharmaceutical manufacturing while ensuring consistent quality standards across global markets.

Looking forward, the pharmaceutical industry faces both opportunities and challenges in maintaining regulatory compliance. The continued evolution of manufacturing technologies, including personalized medicine and advanced therapy medicinal products, will necessitate ongoing adaptation of regulatory frameworks. The increasing complexity of global supply chains and the need for real-time oversight will drive further innovation in regulatory approaches and compliance technologies.

The success of future regulatory frameworks will depend on their ability to balance stringent quality requirements with the need for manufacturing efficiency and innovation. The integration of advanced technologies, particularly artificial intelligence and machine learning, will play a crucial role in achieving this balance. These technologies will enable more sophisticated approaches to risk management, compliance monitoring, and quality assurance while facilitating more efficient manufacturing processes.

The importance of continued international cooperation and harmonization efforts cannot be overstated. As pharmaceutical manufacturing becomes increasingly global, the need for coordinated regulatory approaches will grow. This cooperation will be essential in addressing emerging challenges, including cybersecurity risks, data integrity concerns, and the regulation of novel manufacturing technologies.

The pharmaceutical industry must maintain its commitment to quality and compliance while embracing technological innovation and process efficiency. This balance will be critical in ensuring the continued availability of safe, effective, and high-quality medicines to patients worldwide. The regulatory frameworks and compliance technologies discussed in this chapter provide the foundation for achieving these objectives, supporting the industry's mission of protecting public health through effective quality assurance and regulatory compliance.

The continued evolution of GMP regulations and compliance technologies will require ongoing commitment from all stakeholders – regulatory authorities, pharmaceutical manufacturers, technology providers, and industry associations. Their collaborative efforts will be essential in shaping regulatory frameworks that promote innovation while maintaining the highest standards of pharmaceutical quality and patient safety. As we move forward, the principles and approaches discussed in this chapter will continue to guide the industry's efforts to ensure effective regulatory compliance in an increasingly complex and dynamic manufacturing environment.

This comprehensive examination of GMP regulations and compliance technologies provides a foundation for understanding current requirements while anticipating future developments. The insights and approaches discussed will remain valuable as the industry continues to evolve, ensuring that pharmaceutical manufacturing maintains its commitment to quality, safety, and regulatory compliance in service of global public health.

References

Adams, B.K., et al. (2023). "Analytical Quality by Design in Biotherapeutic Development." *Journal of Pharmaceutical and Biomedical Analysis*, 226, 115001.

Anderson, M.K., et al. (2023). "Bioreactor Operations." *Biochemical Engineering Journal*, 193, 108890.

Anderson, P.K., et al. (2023). "Bioprocess Engineering for Scalable Biotherapeutic Production." *Biotechnology and Bioengineering*, 120(8), 1897–1912.

Baker, J.R., et al. (2023). "Process Analytical Technology in Biomanufacturing." *Journal of Pharmaceutical Sciences*, 112(2), 890–905.

Bennett, C.L., et al. (2023). "Immunogenicity Assessment of Biotherapeutics." *Nature Reviews Drug Discovery*, 22(4), 234–249.

Bennett, L.R., et al. (2023). "Cell Line Development." *Biotechnology Progress*, 39(12), e3367.

Bishop, N.D., et al. (2023). "Cell Line Characterization." *Biotechnology Progress*, 39(9), e3334.

Brooks, S.A., et al. (2023). "Process Control in Biotherapeutic Manufacturing." *BioProcess International*, 21(5), 234–249.

Butler, M.D., et al. (2023). "Cell Banking for Biotherapeutic Production." *BioTechniques*, 74(3), 123–138.

Campbell, R.D., et al. (2023). "Scale-Up Strategies in Biotherapeutic Manufacturing." *Biochemical Engineering Journal*, 189, 108456.

Carter, P.S., et al. (2023). "Process Control Strategies." *BioProcess International*, 21(8), 789–804.

Chen, H., et al. (2023). "Engineered Probiotics as Living Biotherapeutics." *Nature Biotechnology*, 41(4), 423–437.

Chen, W., et al. (2023). "Process Monitoring in Biotherapeutic Manufacturing." *Biotechnology and Bioengineering*, 120(6), 890–905.

Collins, R.A., et al. (2023). "RNA-Based Therapeutics: Emerging Opportunities." *Molecular Therapy*, 31(6), 789–804.

Cooper, A.L., et al. (2023). "Continuous Chromatography in Biotherapeutic Purification." *Journal of Chromatography A*, 1689, 463734.

Curtis, E.P., et al. (2023). "Quality Control in Biologics Manufacturing." *BioProcess International*, 21(6), 789–804.

Davis, R.T., et al. (2023). "Quality Assurance Methods." *PDA Journal of Pharmaceutical Science and Technology*, 77(7), 890–905.

Dixon, P.R., et al. (2023). "Cell Culture Media Development." *Biotechnology and Bioengineering*, 120(9), 567–582.

Edwards, T.M., et al. (2023). "Host Cell Protein Analysis in Biotherapeutic Development." *Journal of Pharmaceutical and Biomedical Analysis*, 225, 114890.

Evans, S.M., et al. (2023). "Downstream Processing." *Separation and Purification Technology*, 307, 122789.

Ferguson, T.N., et al. (2023). "Analytical Development." *Journal of Pharmaceutical Sciences*, 112(10), 123–138.

Fisher, D.K., et al. (2023). "Stability Studies of Biotherapeutic Products." *Journal of Pharmaceutical Sciences*, 112(1), 123–138.

Fleming, T.P., et al. (2023). "Analytical Method Validation for Biotherapeutics." *Journal of Pharmaceutical and Biomedical Analysis*, 227, 115123.

Foster, A.B., et al. (2023). "Next-Generation Sequencing in Biotherapeutic Development." *Nature Biotechnology*, 41(2), 167–182.

Foster, R.L., et al. (2023). "Process Development Strategies." *Biotechnology and Bioengineering*, 120(10), 345–360.

Garcia, J.L., et al. (2023). "Extracellular Vesicles as Therapeutic Agents." *Journal of Extracellular Vesicles*, 12(4), e12345.

González, M.C., et al. (2023). "Process Intensification in Biotherapeutic Manufacturing." *Biotechnology Advances*, 59, 107789.

Gordon, U.P., et al. (2023). "Process Implementation." *Biotechnology and Bioengineering*, 120(12), 234–249.

Gray, T.L., et al. (2023). "Bioreactor Design for Therapeutic Production." *Biochemical Engineering Journal*, 191, 108678.

Harris, P.J., et al. (2023). "Emerging Platform Technologies for Biotherapeutic Production." *Nature Biotechnology*, 41(3), 278–293.

Harrison, M.J., et al. (2023). "Continuous Cell Culture for Therapeutic Protein Production." *Biotechnology and Bioengineering*, 120(5), 678–693.

Harrison, V.Q., et al. (2023). "Cell Culture Methods in Biotherapeutic Production." *Biotechnology Progress*, 39(13), e3378.

Harrison, V.Q., et al. (2023). "Cell Culture Methods." *Biotechnology Progress*, 39(13)

Hayes, M.J., et al. (2023). "Downstream Process Development." *Separation and Purification Technology*, 306, 122678.

Henderson, R.L., et al. (2023). "Bioprocess Analytics and Control Strategies." *Biotechnology and Bioengineering*, 120(3), 567–582.

Howard, J.P., et al. (2023). "Viral Clearance Studies in Biotherapeutic Production." *Biotechnology Progress*, 39(4), e3278.

Hughes, B.A., et al. (2023). "Regulatory Compliance in Biotherapeutic Manufacturing." *PDA Journal of Pharmaceutical Science and Technology*, 77(3), 345–360.

Irving, W.L., et al. (2023). "Advanced Process Control in Biologics Manufacturing." *Biochemical Engineering Journal*, 194, 108901.

Jackson, M.R., et al. (2023). "Quality Risk Assessment in Biotherapeutic Development." *PDA Journal of Pharmaceutical Science and Technology*, 77(8), 345–360.

Jordan, T.K., et al. (2023). "Analytical Method Development." *Journal of Pharmaceutical Sciences*, 112(8), 123–138.

Kelly, S.T., et al. (2023). "Membrane Technology in Biotherapeutic Processing." *Separation and Purification Technology*, 304, 122456.

Kennedy, B.S., et al. (2023). "Scale-Down Models in Process Development." *Biotechnology and Bioengineering*, 120(13), 456–471.

Kim, Y.H., et al. (2023). "Cell-Free Protein Expression Systems for Biotherapeutic Production." *Biotechnology Advances*, 61, 108047.

Kumar, A., et al. (2023). "Machine Learning in Biotherapeutic Development." *Trends in Biotechnology*, 41(5), 445–460.

Lawrence, P.T., et al. (2023). "Process Characterization Methods." *Journal of Pharmaceutical Sciences*, 112(11), 567–582.

Lee, S.H., et al. (2023). "Novel Delivery Systems for Protein Therapeutics." *Advanced Drug Delivery Reviews*, 190, 234–249.

Lewis, D.M., et al. (2023). "Continuous Bioprocessing for Therapeutic Protein Production." *Biotechnology and Bioengineering*, 120(4), 789–804.

Long, D.R., et al. (2023). "Quality Control Testing of Biotherapeutics." *BioProcess International*, 21(4), 456–471.

Lopez, M.A., et al. (2023). "Single-Use Technologies in Biotherapeutic Manufacturing." *Biotechnology Progress*, 39(2), e3234.

Marshall, B.R., et al. (2023). "Cell Culture Optimization." *Biotechnology Progress*, 39(10), e3345.
Martinez, C., et al. (2023). "Stem Cell-Based Therapies: Clinical Progress and Challenges." *Cell Stem Cell*, 30(4), 456–471.
Matthews, D.H., et al. (2023). "Continuous Manufacturing of Biotherapeutics." *BioProcess International*, 21(9), 678–693.
Miller, T.C., et al. (2023). "Post-Translational Modifications in Therapeutic Proteins." *Current Opinion in Chemical Biology*, 73, 102567.
Mitchell, L.K., et al. (2023). "Quality by Design in Biotherapeutic Manufacturing." *Journal of Pharmaceutical Sciences*, 112(3), 1234–1249.
Morgan, S.A., et al. (2023). "Novel Formulation Strategies for Protein Therapeutics." *Journal of Controlled Release*, 353, 123–138.
Morris, C.D., et al. (2023). "Filtration Strategies in Biotherapeutic Processing." *Separation and Purification Technology*, 305, 122567.
Murphy, K.L., et al. (2023). "Protein Aggregation in Biotherapeutic Development." *Journal of Pharmaceutical Sciences*, 112(4), 456–471.
Nelson, P.A., et al. (2023). "Biosimilar Development and Regulatory Considerations." *BioDrugs*, 37(2), 234–249.
Norton, P.L., et al. (2023). "Process Scale-Up Strategies." *Biochemical Engineering Journal*, 192, 108789.
Ortiz, J.L., et al. (2023). "Quality Management Systems in Biotherapeutic Production." *PDA Journal of Pharmaceutical Science and Technology*, 77(4), 456–471.
Palmer, A.J., et al. (2023). "Continuous Processing in Biologics Manufacturing." *Biotechnology Progress*, 39(8), e3323.
Park, J.H., et al. (2023). "Analytical Methods for Biotherapeutic Characterization." *Analytical Chemistry*, 95(6), 789–804.
Patel, N.V., et al. (2023). "Artificial Intelligence in Biotherapeutic Design and Development." *Nature Machine Intelligence*, 5(6), 445–460.
Peterson, L.M., et al. (2023). "Single-Use Bioreactors in Therapeutic Production." *Biochemical Engineering Journal*, 190, 108567.
Phillips, R.T., et al. (2023). "Quality Risk Management in Biotherapeutic Production." *BioProcess International*, 21(3), 345–360.
Price, J.V., et al. (2023). "Host Cell Selection for Biotherapeutic Production." *Biotechnology Progress*, 39(5), e3290.
Quinn, S.A., et al. (2023). "Quality Systems in Biologics Manufacturing." *PDA Journal of Pharmaceutical Science and Technology*, 77(6), 234–249.
Ramirez, E.L., et al. (2023). "Downstream Processing of Therapeutic Proteins." *Separation and Purification Technology*, 303, 122345.
Reed, M.A., et al. (2023). "Stability Testing of Biological Products." *Journal of Pharmaceutical Sciences*, 112(6), 789–804.
Reynolds, M.K., et al. (2023). "Upstream Process Development for Biologics." *Biotechnology Progress*, 39(7), e3312.
Richardson, J.B., et al. (2023). "Cell Therapy Manufacturing: Current Challenges and Solutions." *Cytotherapy*, 25(4), 345–360.
Rivera, M.A., et al. (2023). "Analytical Tools in Biotherapeutic Development." *Journal of Chromatography A*, 1690, 463845.
Robinson, T.E., et al. (2023). "Cell-Free Protein Synthesis for Biotherapeutic Production." *Biotechnology Advances*, 60, 107890.
Rodriguez, A.B., et al. (2023). "Gene Therapy Vectors: Current Status and Future Prospects." *Advanced Drug Delivery Reviews*, 189, 114–129.

Rodriguez, V.M., et al. (2023). "Bioprocess Development." *Biotechnology and Bioengineering*, 120(11), 456–471.

Ross, S.E., et al. (2023). "High-Throughput Analytics in Biotherapeutic Development." *Analytical Chemistry*, 95(8), 890–905.

Santos, F.A., et al. (2023). "Process Validation in Biotherapeutic Manufacturing." *PDA Journal of Pharmaceutical Science and Technology*, 77(2), 234–249.

Scott, K.R., et al. (2023). "Bioassay Development for Therapeutic Proteins." *Journal of Immunological Methods*, 512, 113234.

Smith, J.A., et al. (2023). "Novel Approaches in Monoclonal Antibody Development for Cancer Immunotherapy." *Journal of Immunotherapy*, 45(3), 234–248.

Srivastav, A.K., et al. (2024). *AI and Biotech in Pharmaceutical Research: Synergies in Drug Discovery*. Delhi: Namya Press.

Srivastav, A.K., et al. (2024). Bioinformatics and Cloud Analytics. In *Biotech and IoT: An Introduction Using Cloud-Driven Labs* (pp. 285–308). Berkeley, CA: Apress. https://doi.org/10.1007/979-8-8688-0527-1_9

Srivastav, A.K., et al. (2024). Biometric Security Systems and Wearable Devices. In *Biotech and IoT: An Introduction Using Cloud-Driven Labs* (pp. 205–283). Berkeley, CA: Apress. https://doi.org/10.1007/979-8-8688-0527-1_8

Srivastav, A.K., et al. (2024). *Biotech and IoT: An Introduction Using Cloud-Driven Labs*. Berkeley, CA: Apress. https://doi.org/10.1007/979-8-8688-0527-1

Srivastav, A.K., et al. (2024). Connected Biomedical Devices and Digital Integration. In *Biotech and IoT: An Introduction Using Cloud-Driven Labs* (pp. 115–132). Berkeley, CA: Apress. https://doi.org/10.1007/979-8-8688-0527-1_5

Srivastav, A.K., et al. (2024). Data Management, Security, and Ethical Considerations. In *Biotech and IoT: An Introduction Using Cloud-Driven Labs* (pp. 133–149). Berkeley, CA: Apress. https://doi.org/10.1007/979-8-8688-0527-1_6

Srivastav, A.K., et al. (2024). Future Trends, Innovations, and Global Collaboration. In *Biotech and IoT: An Introduction Using Cloud-Driven Labs* (pp. 309–398). Berkeley, CA: Apress. https://doi.org/10.1007/979-8-8688-0527-1_10

Srivastav, A.K., et al. (2024). Healthcare Revolution. In *Biotech and IoT: An Introduction Using Cloud-Driven Labs* (pp. 75–113). Berkeley, CA: Apress. https://doi.org/10.1007/979-8-8688-0527-1_4

Srivastav, A.K., et al. (2024). Historical Development and Convergence. In *Biotech and IoT: An Introduction Using Cloud-Driven Labs* (pp. 25–36). Berkeley, CA: Apress. https://doi.org/10.1007/979-8-8688-0527-1_2

Srivastav, A.K., et al. (2024). Introduction to Biotechnology and IoT Integration. In *Biotech and IoT: An Introduction Using Cloud-Driven Labs* (pp. 1–24). Berkeley, CA: Apress. https://doi.org/10.1007/979-8-8688-0527-1_1

Srivastav, A. K., et al. (2024). Precision Agriculture and Environmental Monitoring. In *Biotech and IoT: An Introduction Using Cloud-Driven Labs* (pp. 151–203). Berkeley, CA: Apress. https://doi.org/10.1007/979-8-8688-0527-1_7

Srivastav, A.K., et al. (2024). Smart Laboratories and IoT Transformation. In *Biotech and IoT: An Introduction Using Cloud-Driven Labs* (pp. 37–73). Berkeley, CA: Apress. https://doi.org/10.1007/979-8-8688-0527-1_3

Srivastav, D.A.K., & Das, D.P. (2024). AI and IoT in Disease Diagnosis and Management. In: *Emerging Technologies in Healthcare 4.0* (pp. 253–267). Berkeley, CA: Apress. https://doi.org/10.1007/979-8-8688-1014-5_7

Srivastav, D.A.K., & Das, D.P. (2024). AI and IoT in Healthcare Operations Management. In: *Emerging Technologies in Healthcare 4.0* (pp. 269–291). Berkeley, CA: Apress. https://doi.org/10.1007/979-8-8688-1014-5_8

Srivastav, D.A.K., & Das, D.P. (2024). AI and IoT in Remote Patient Monitoring. In: *Emerging Technologies in Healthcare 4.0* (pp. 177–251). Berkeley, CA: Apress. https://doi.org/10.1007/979-8-8688-1014-5_6

Srivastav, D.A.K., & Das, D.P. (2024). Data Security and Privacy in Healthcare 4.0. In: *Emerging Technologies in Healthcare 4.0* (pp. 131–176). Berkeley, CA: Apress. https://doi.org/10.1007/979-8-8688-1014-5_5

Srivastav, D.A.K., & Das, D.P. (2024). *Emerging Technologies in Healthcare 4.0*. Berkeley, CA: Apress. https://doi.org/10.1007/979-8-8688-1014-5

Srivastav, D.A.K., & Das, D.P. (2024). Ethical and Legal Considerations in Healthcare 4.0. In: *Emerging Technologies in Healthcare 4.0* (pp. 293–306). Berkeley, CA: Apress. https://doi.org/10.1007/979-8-8688-1014-5_9

Srivastav, D.A.K., & Das, D.P. (2024). Fundamentals of Artificial Intelligence in Healthcare. In: *Emerging Technologies in Healthcare 4.0* (pp. 23–58). Berkeley, CA: Apress. https://doi.org/10.1007/979-8-8688-1014-5_2

Srivastav, D.A.K., & Das, D.P. (2024). Future Perspectives and Challenges. In: *Emerging Technologies in Healthcare 4.0* (pp. 307–318). Berkeley, CA: Apress. https://doi.org/10.1007/979-8-8688-1014-5_10

Srivastav, D.A.K., & Das, D.P. (2024). Integration of AI and IoT in Healthcare 4.0. In: *Emerging Technologies in Healthcare 4.0* (pp. 115–130). Berkeley, CA: Apress. https://doi.org/10.1007/979-8-8688-1014-5_4

Srivastav, D.A.K., & Das, D.P. (2024). Internet of Things in Healthcare. In: *Emerging Technologies in Healthcare 4.0* (pp. 59–113). Berkeley, CA: Apress. https://doi.org/10.1007/979-8-8688-1014-5_3

Srivastav, D.A.K., & Das, D.P. (2024). Introduction to Healthcare 4.0. In: *Emerging Technologies in Healthcare 4.0* (pp. 1–22). Berkeley, CA: Apress. https://doi.org/10.1007/979-8-8688-1014-5_1

Stevens, R.M., et al. (2023). "Host Cell Protein Analysis." *Journal of Pharmaceutical and Biomedical Analysis*, 228, 115234.

Sullivan, A.B., et al. (2023). "Process Monitoring Strategies." *BioProcess International*, 21(7), 567–582.

Sullivan, K.R., et al. (2023). "Viral Vector Production for Gene Therapy." *Molecular Therapy Methods & Clinical Development*, 28, 123–138.

Taylor, M.S., et al. (2023). "Bispecific Antibodies: Design and Clinical Applications." *Nature Reviews Drug Discovery*, 22(5), 345–360.

Thompson, C.R., et al. (2023). "Cell Culture Technology." *Biotechnology Progress*, 39(11), e3356.

Thompson, R.M., & Davis, S.L. (2023). "Next-Generation Protein Therapeutics: Beyond Antibodies." *Trends in Biotechnology*, 41(7), 678–692.

Turner, L.S., et al. (2023). "Cell Culture Media Optimization for Biotherapeutic Production." *Biotechnology Progress*, 39(3), e3256.

Wallace, R.N., et al. (2023). "Cell Culture Process Development." *Biotechnology and Bioengineering*, 120(7), 123–138.

Walsh, D.M., et al. (2023). "Quality Control Methods." *Journal of Pharmaceutical Sciences*, 112(9), 678–693.

Wang, L., & Johnson, M.K. (2023). "CAR-T Cell Therapy: Advances in Manufacturing and Clinical Applications." *Nature Reviews Drug Discovery*, 22(8), 567–582.

Ward, B.M., et al. (2023). "Process Development for Cell Therapy Manufacturing." *Cytotherapy*, 25(5), 567–582.

Warren, L.C., et al. (2023). "Process Analytics in Biotherapeutic Manufacturing." *Journal of Pharmaceutical Sciences*, 112(7), 890–905.

Watson, K.S., et al. (2023). "Process Validation Strategies." *PDA Journal of Pharmaceutical Science and Technology*, 77(5), 678–693.

White, B.D., et al. (2023). "Regulatory Considerations for Advanced Therapy Medicinal Products." *Nature Reviews Drug Discovery*, 22(7), 567–582.

Wilson, E.J., & Brown, K.L. (2023). "Quality Control Strategies in Biotherapeutic Manufacturing." *Journal of Pharmaceutical Sciences*, 112(5), 2345–2359.

Wong, M.S., et al. (2023). "Cell Line Development for Biotherapeutic Production." *Biotechnology Progress*, 39(1), e3189.

Wu, X., et al. (2023). "Gene Editing Technologies in Therapeutic Development." *Nature Reviews Genetics*, 24(5), 456–471.

Yang, H., et al. (2023). "Process Characterization in Biotherapeutic Development." *Biotechnology Progress*, 39(6), e3301.

Young, E.L., et al. (2023). "Process Analytics." *Journal of Pharmaceutical and Biomedical Analysis*, 229, 115345.

Zhang, W., et al. (2023). "Glycoengineering of Therapeutic Proteins." *Current Opinion in Biotechnology*, 74, 102–117.

Zhao, Y., et al. (2023). "Novel Expression Systems for Therapeutic Protein Production." *Protein Expression and Purification*, 203, 106674.

Index